Thinking Outside The Soil

How Hydroponic Fodder Helps Farmers
Save Water, Improve Livestock Quality, and
Become Better Stewards

Sean Short

Blooming Scribe Media

Dedication:

To my brother Kyle, whose spirit and joy for life will always be a beacon on my path of stewardship and innovation. Your love is missed but forever echoes in the spaces where we once dreamed together.

And to Elliott, one of the original Chicken Pimps whose hands-on dedication and care for our feathered friends laid the foundation for the principles this book advocates. Your wisdom in the coop has transcended into the wisdom on these pages.

May this work honor your memories by sowing the seeds of change and nurturing growth—just as you both did in life. In every root that finds water and every chick that finds strength, your influences live on, reminding us that the essence of our work is love and compassion for all.

Blooming Health Farms

Growing Food. Growing People.

Blooming Health Farms is a 501c3 nonprofit aquaponics and chicken egg-laying farm in Northern Colorado that helps and employs at-risk youth between the ages of 15 and 24 who are involved with the criminal justice system. We get kids off the street by empowering them with job skills, STEM education, and mental health mentorship in order to create contributing members of our community. Please visit https://www.bloominghealthfarms.com to learn how you could help us transform the Northern Colorado community.

As part of our empowerment model, we incentivize our at-risk youth with commission sales on eggs and products we make. The at-risk youth do the work needed to take care of our aquaponics and chickens. We grow leafy herbs, our own feed and better than organic eggs. We also make the things needed to help farmers do it themselves. That mostly means we make simple kits for farmers so they too can grow their own poultry feed without a lot of work or fuss. We call these our Sprouting for Success Kits

We make a few versions from small to large. The one featured here is the one we use to teach the principles of sprouting to kids, community members and farmers. It comes with a welded stand and we hold a welding class to teach simple welding techniques. We also make the 3D printed lid that fits on any wide-mouth mason jar, and show kids the technical aspects of printing and programming. The goal is to provide skills we know are valuable for employment. Sales, Procurement, Welding and 3D printings are just a some examples.

"The more [we] sell the more youth I get to help"
Sean Short

What Sets Me Apart

A consultant or engineer delivers, and then the user is left hanging. Many companies can bridge that gap with excellent customer service. However, it's imperative that something is developed with you as a farmer. Every system needs time to settle into its unique environment. There will always be subtleties that don't follow suit and deviate from the set instructions.

I take the time to assess your needs and set you up for as much success as possible. That could mean writing a book to explain a new topic. Or it may mean tweaking an established process you're using, but that may not work. Process improvement is important.

Especially when time is money. And time is very precious while running or working on the farm.

I approach all challenges with a needs analysis. This generally begins with some sort of operational deficiency. Water and land are scarce. Money is tight. It may also begin with some technological opportunity. Hydroponic systems save resources. Systems increases your bottom line.

This is best captured on an 'Initial Project Needs Assessment.' The assessment gathers basic information about the purpose and scope of a project. It is short and allows for a quick consult, without wasting anyone's time. This is where we would talk about if we are a good fit and next steps. Ideally, I would want to advise you through at least one crop cycle.

I try to make sure people have the best information so they can make informed decisions. Years ago I worked for a company where I did System Sales, Design and Product Support. Part of that job was to tell customers accurate information so they could make a sound investment.

I often talked people out of purchasing aquaponics systems until they got better experience. Many people wanted to pursue aquaponics because it was new.

I recommend simple systems so people could learn hydroponic principles without a huge investment. Learning is expensive and we must use our time and resources wisely.

A hydroponic project can differ in the time needed for completion. Some systems are "off-the-shelf" and minimal time is required for planning. Setup can be done in a few days or less.

Extensive projects may take weeks or months to complete and need proper planning. These projects need systems engineering consultation to deliver a workable concept. Design and implementation of a solution. What sets me apart is my commitment to a process that is both helpful and useful. I would love to hear from anyone who wants to pursue a hydroponic endeavor (excerpt from *Chapter 1*).

Visit https://www.thinkingoutsidethesoil.comor scan the QR code to find your free questionnaire. Check it out and set up a complimentary 20-minute consultation when you are ready.

Alternatively, email sean@thinkingoutsidethesoil.com with any questions.

At Blooming Health Farms, we champion a new breed of farmer—the Chicken Pimp and the Chicken Peep.

Let Us Explain...

We've reframed 'pimp' to symbolize our innovative approach to farming. It means to make things better. Because here, Pimpin' means pushing boundaries and driving change.

While **"*Chicken Pimp*"** encapsulates the audacious spirit of our movement, we use the term **"*Chicken Peep*."** for those sprouting with potential taking the first step on the journey towards becoming a "Pimp."

Our manifesto guides our journey as stewards of the land and nurturers of both chickens and change.

The *Chicken Pimp* Manifesto:

- A *Pimp* leads with humility, overcomes challenges, gets things done, and demands the highest quality for the flock.

- A *Pimp* negotiates agricultural knowledge and educates others. A chicken pimp knows a balance between imparting wisdom and learning from nature.

- A *Pimp* is savvy beyond business. Pimps harness innovative, sustainable methods and push the boundaries of traditional chicken farming.

- A *Pimp* is influential within the community and creates a platform to connect people with their food source.

- And...a *Pimp* is loving to all. Compassion demonstrates a willingness to follow the strictest welfare and quality standards. A Chicken Pimp knows that **love** leads to a resilient food system.

The craft of **Chicken Pimpin'** is more than a pursuit; it's a rebellion against current chickening. A radical reimagining of our agricultural landscape.

At **Blooming Health Farms**, we're not just **growing food** and **growing people**. We're also **sowing the seeds of change in the chicken world.**

You are the heart of this *revolution*...**Welcome to our Flock!**

Contents

List of Photo Credits

1. *Satellite view of the Gulf. NASA-NOAH image.* "How Our Food System Is Destroying the Nation's Most Important Fishery." Grist, May 18, 2012. https://grist.org/article/2010-02-08-who-owns-the-dead-zone/.

2. *Tomato Tree. Epcot Center.* https://upload.wikimedia.org/wikipedia/commons/a/aa/Tomatotree.JPG.

3. Torres, David E. *Artist Depiction of The Hanging Gardens of Babylon.* HYDROPONICS: THE ART OF GROWING PLANTS WITHOUT SOIL. Accessed February 2, 2021. https://www.landuum.com/en/history-and-culture/hydroponics-the-art-of-growing-plants-without-soil/.

4. *Egyptian Hieroglyph.* Egypt and Aquaculture/Hydroponics. Institute of Simplified Hydroponics. Accessed February 2, 2021. http://www.carbon.org/school/newclass/egypt.htm.

5. *Aztec Chinampas model by Te Mahi, Photographer: Te Papa.* https://www.pinterest.nz/pin/749075350490894339/.

6. *W.F. Gericke and his wife with a 12 ft. tall hydroponic tomato.* http://hydroponicgardening.com/history-of-hydroponics/the-birth-of-hydroponics/.

ek.wordpress.com/2013/09/05/fall_for_fodder/.

16. Julie Hanscome. *Goats eating barley at Hanscome Dairy.* 2013. https://farmtek.wordpress.com/2013/05/28/spotlight-hanscome-dairy/.

17. *Pig eating barley.* 2013. https://farmtek.wordpress.com/2013/01/08/abigails-fodder-for-thought-use-what-youve-got/.

18. ClearSpan Fabric Structure. *Alpaca eating barley.* https://farmtek.wordpress.com/2013/07/29/fodder_for_alpacas/.

19. Hawaii News Now. *Sean Short and Dr. Harry Ako tend to an aquaponics system.* 2012. https://www.hawaiinewsnow.com/story/16952337/backyard-aquaponic-system-produces-green-greens/.

20. Sean Short. *BHF Sprouting Kit.* 2022.

21. Sean Short. *Two dry ounces of barley seed.* 2022.

22. Sean Short. *Two dry ounces of barley seed in jar.* 2022.

23. Sean Short. *Barley with dirty rinse water.* 2022.

24. Sean Short. *Initial barley seed soak.* 2022.

25. Sean Short. *Soaked barley seed after 8 hours.* 2022.

26. Sean Short. 26. *Rinsed barley seed on rack to drain.* 2022

Preface to the Original Edition

Work on this book began in early June of 2022 as a challenge to write a book in thirty days based on my experience with hydroponics. I was soon inundated with calls about farmers going out of business due to water-related issues. But farmers wanted more than just water; they needed feed. Fields are drying up, and farmers can't get hay. Repeated drought makes it worse each year. I wanted to write a book that helps farmers find another way to maintain their livelihoods with what I knew.

I didn't know the depth of what was required. Lots of coffee, conversations, written proposals, and sleepless nights snuggled up with research papers. Over 120 articles and ten textbooks in less than four months. All but a few were new.

Hydroponic fodder has assumed such enormous dimensions that a popular, practical, and non-technical text dealing with these methods is urgently needed. Yet, this is just a drop in the bucket of what can be said about possible solutions to drought. More work is required from farmers, researchers, policymakers, and others like myself.

Preface to the Revised Edition

A year since the original was released. And never did I expect to become a chicken man and find a path with more youth that resonated with chickens. As a small-scale egg laying operation, we have had more success that I originally expected and we have been called to serve a sub-population of people that need the help as much as the at risk youth.

But it was one youth in particular that has truly shaped who we are and where we are going. He came to me from a local teacher. One that moonlighted as a delivery driver. The intersection of at-risk youth and mental health wasn't truly apparent until Kewani came to Alpha site one day and ate a tomato. Ah the giant Epcot tomato that captured me lived on in the form of a cherry tomato. One I bred that taste as sweet as a starburst candy. Kewani had never had a fresh tomato before and after his eyes lit up, I knew I had him for life. But of course it was what we did together that truly solidified our bond as teacher and student. He wants to be a rapper. And I immediately fueled his dream. Soon I saw a delinquent become a caring and compassionate kid towards the chickens that we had. Without any encouragement, Kewani came to love and care for our feathered friends.

I took him to the farmers market and then there came the greatest three sentence business plan I had heard in my life after we sold out of eggs in the first fifteen minutes. "We need more chickens." "We can't," I replied...Kewani persisted with his chicken math, "Yeah, but Sean, if we had just two more chickens, we could get two more eggs a day and bring more eggs to the next market. That means we could make more money."

The idea was brilliant. And a bit reckless what happened next. But the Urban Farming concept took on a whole new level. Never would anyone believe that nestled in the heart of an agricultural haven, one that has an odd ordinance against chickens, that we would become a chicken farm.

I found a home for them on the edge of the city after meeting a couple that wanted to farm. You see the second group of people we found that needed our help are the homesteaders that were hurting and struggling to be profitable.

We made a pact and soon pimped their property in to our first co-op and egg laying operation. And boy, was it rough at first. We quickly realized that there is a reason it is hard to be a profitable farmer. Enter our sprouts. In the past year we have affectionally called our sprouts 'Chicken Crack.' And it's so easy to grow that Ethan took to it within a week.

We're now pursuing the Certified Humane process and are growing "better than organic" eggs. What does better than organic mean though? It means that we are adhering to the strictest organic and animal welfare standards while still being as frugal as farmers can be.

Acknowledgments

This book was a lot of fun to write. Little did I know it would be others that helped shape the voice of this book.

I want to thank the farmers on the Weld Extension Council. Your support and advocacy showed me that we need different ways to feed our livestock. Within minutes of suggesting I write a book, I had three phone calls from excited farmers saying, "I love what you are doing." So without that, there would be no book. As it is for you, the rancher and farmers struggling to find their way in a resource-strapped world. Especially one where your water is disappearing.

To my father, I would like to thank him for his years of being who he was, a veterinarian who believed in public safety and service. I learned a lot about cattle from him while I was a kid. Today, I thank him for proofing my work when he could.

I thank Joshua Sprague, Ray Brehm, Paul G. Broadie, TJ Scott, and Ryan Smith. These people helped me write and, as talented people in other areas, were integral to my success. You all helped me write and go through the publishing process. It is a lot of work, so thank you for the advice, virtual hand-holding, late-night emails or calls, and struggles that made me a better author.

I am grateful and inspired by JD and Tawnya Sawyer for introducing me and enamoring me with aquaponics. I may have found aquaponics without them, yet I am ok with believing it's all their fault. I likely wouldn't have rushed off to Hawai'i if I didn't meet JD. I had the pleasure of working for them many years ago, and they have been trusted friends and mentors ever since.

Hands down, I attribute much of my education and ability to think like I do today to Dr. Harry Ako. However, it took him almost three years to break me. He often called me a cowboy, and I was as stubborn as one. It wasn't until I won a research award that I understood why I was taught the way I was. I'm grateful to have an education from some leaders in their schools of thought, like Dr. Harry Ako, Ed Otsuji, Dr. Kent Kobayashi, Dr. Mark Merlin, Noel Huber, Steven M. Biemer, and Dr. Judith Dahmann.

No project happens without others being on the sidelines. So, I would like to thank those who have encouraged me while working on this project. Or those whose recent involvement helped me get to a point where I could write a book. Thank you, Mom, Ed VanDyne, Tim Hubbard, John and Mary Gauthiere, Frankie, Kaiden, TJ, Kewani, Rachel, Jim K., John von Tungeln, Robbie, Jason, Mike, Sarah, Joelene, Misty, Courtnay, Samantha, Megan, Anthony, Dale, Junior, Jesse, Kendal, Dolly, Ryan, Cheryl, Josh Hoban, Brandon, Ulysses, Elliott, Macoy, Nadia, as well as Lindsay, Isaiah, and Steve.

I also study successful ones to model them rather than reinvent the wheel. For the work you have done and immortalized for others, I would like to thank Luther Burbank, Nikola Tesla, Mark Twain, P.T. Barnum, Albert Hofmann, Temple Grandin, Bruce

Lee, Mary Kay, Napoleon Hill, R. Buckminister Fuller, Terence McKenna, Marie Curie, Napoleon Hill, Temple Grandin.

And finally, thank you to those who have read my work and all I have met or will meet.

Greeley, Colorado
October 2022

Chapter One

EXTENSION

Extension (n) - The application of scientific research and new knowledge to agricultural practices through [farmer] education.

We cannot fight the existing model by trying to change it, but we can introduce a new model that makes the old model obsolete.

R. Buckminister Fuller

E ach person's head darts to the left as they enter the conference room. A table next to the door is adorned with a stack of small brown boxes. An address sized label reads, `Turkey`, `Ham`, or `Roast Beef`. Everyone who enters grabs a box, napkin, and bottled water from an iced cooler. I grab a napkin and smile as I reach for the box marked `Roast Beef`.

Inside the cardboard container are edible delights. There is a brownie, a bag of Lays potato chips, a pickle, and thankfully, a Roast Beef Sandwich. I unscrew my water bottle and take a swig before opening my brownie. I make sure to eat dessert first. That way, you know you get it.

Idle chatter begins as more and more people enter the room. Then, our Chair, Amy, enters with a couple of ladies while laughing. I wonder to myself, "How come *she* looks so familiar?" The three ladies finally sit with their lunch, begin to open the box and crack open the bottled water.

"Welcome," Amy says as she sets down her food and water, "should we call this meeting to order?" The murmur stopped, and only the sounds of food wrappers broke the silence. She continued, "Thank you for being here today. For those of you who may not know, Hannah is the new Director of Extension here in Weld. Hannah, would you like to kick it off?"

The guy named Gary to my right unwraps his Turkey sandwich and takes a bite. He grabs a pen from his shirt pocket and signs the piece of paper on the clipboard. He squiggles loudly before the board gets to me without the pen. I carry my own pen as well.

"Has everyone signed in?" asks Hannah. "Good. First, I would like to congratulate the new members of the Public Advisory Council. Welcome, we're excited to have you! I would also like to congratulate Gary and Sean on their reappointment to the Council. Thank you for what you do."

She continues, "Well since we have a bunch of new faces, I think it would be good to go around the room and introduce ourselves again. Let's say who we are and tell us something interesting about you. Maybe about what you do or something fun going on."

The County Commissioner began, "Hi, I'm Perry Buck, and I am just thrilled to be here. I love ag, and Weld has the best ag in the world. I am here to help you guys in any way!" Her energy for agriculture is impressive. But, of course, so is everyone else's energy. "I'm Kat, the new agronomist for Weld and Washington Counties. I moved here from Nebraska, where I was the Agronomist in McCook County." Young like me, she would soon help shape the destiny of my business and become an unexpected ally.

Three more people introduce themselves as I unwrap my Roast Beef. I squeezed the mayonnaise in a squiggly line on both sides of the bread. I recognized Cindy and Michelle when they entered with our Chairwoman. They have been on the Extension staff since I got here in 2016. Then again, the lady next to Amy has been on Council a while, and I cannot remember her name. Names are my thing.

"I'm Lori. My husband and I own Top Notch Meats and sell at the local markets. Well, for now..." she trailed off as I blush and realized who she was. We smile at each other from across the Greeley Farmers Market on some Saturdays. She and her husband constantly reach in and out of large coolers. Out came a frozen jewel of redness, followed by a smile on each person's face. The coolers contained various cuts of all-natural beef and lamb. Top Notch Meats has been around for some time and produces high-quality products.

Top Notch Meats is located about twenty minutes North of Greeley in a tiny place known as Pierce. Pierce sits on the High Prairie of the Front Range. An area that is windy and dry. The region has experienced an unprecedented drought for over thirty years. They are respected by a lot of the community. Lori is also a

4-H leader here in Northern Colorado. She tells the Council there is no water for her cattle and they might need to cull.

Horticulture position

I swallow my bite after Lori's brief introduction and prepare to introduce myself after Gary. "I am Sean Short. I fill the Horticulture position. I started a nonprofit farm that uses fish water to grow plants. We use only ten percent of the water and a fraction of the space compared to soil agriculture. It's nonprofit because we empower at-risk youth with STEM skills and give them mental health support. All through the medium of hydroponics. We call it Blooming Health Farms"

Many Coloradoans are unfamiliar with hydroponic forms of agriculture unless someone grows cannabis. Old-time farmers know cannabis as a utilitarian crop. Years ago, a coworker told me, "You call it cannabis today. But back in the 20s, my granddad had a setting on his Farmall tractor for hempseed. They used it for feeding cattle and birds." And it's true, hempseed is a significant constituent of some wild bird seeds, and a source of animal feeds in Canada, Italy, and China.

Colorado is a leader in the hydroponic cannabis industry. So when I mentioned the word hydroponics, I saw eyes roll and heard a couple snickers from the newer members. But I also knew that sometimes the most unlikely ideas can seed revolutions so I thought, "Should I tell them?"

What this book is NOT

This book is not a complete guide on how to grow plants, or pot, with hydroponics. Thousands of books and websites can tell you how to grow with various methods, techniques, and processes. A simple search on Google will result in over seventy million hits. Nor is this a complete grow your own animal fodder. That takes a dedicated technical manual, which can supplement your fodder journey.

Those resources may be helpful in your journey, and some are referenced in the back, yet that is out of the scope of this book. It would likely take many years, bore you to death or make you fall asleep at the wheel while listening to the book. I do not want the blame for crashes or curvy rows of crops.

What this book IS

There could be many reasons why you picked up this book. But since everyone needs to eat, there is good reason to keep reading. This book helps farmers and policymakers understand how hydroponic practices save water, optimize livestock quality and promote better stewardship.

The first Chapter talks about the benefits of hydroponic fodder using over ninety percent less water and space than the same crops grown in the field, with the same yields. The chapter then points out the pollution cutback and carbon capture capabilities. Hydroponics' water-saving and land-sparing potential are integral to any climate change discussion.

Chapter Two highlights the origin of hydroponic agriculture as it applies to working with farmers. The chapter continues with the benefits of hydroponic fodder cultivation and how it uses less water and space than the same crops grown in the field.

Chapter Three highlights hydroponic fodder's use among different livestock raised to provide resources like eggs, milk, wool, and feathers. An early source during the course of my research was a visual display of livestock production around the world called StoryMaps.[1] It shows the most raised animals in the world and where they are on earth. I was intrigued to learn that "the combined weight of cattle, chickens, and pigs exceeds the weight of all wild animals and humans combined." The chapter will focus on chickens, horses, cattle, ducks, sheep, goats, pigs. In addition, it will touch on other areas where hydroponic fodder could benefit agriculture like seed production and exotic animals.

Chapter Four is called the ABC'S of Hydroponic Fodder^{TM}, covering common fodder crops. Many commonly grown field crops or weeds that farmers use to feed livestock can be used for hydroponic fodder. However, some crops do better than others. Selecting the best forage crop is important to produce the highest quality fodder. So, we will limit it to seven major crops - alfalfa, barley, corn, cowpea, clover, duckweed, and sunflower. Factors discussed will be availability, water use efficiency, yield, and nutrition.

Chapter Five states the work involved with hydroponic fodder. It covers what you may need, fodder systems mentioned throughout this book, and ways to access additional information so you can be successful. The goal is to give you enough to make an informed decision and ask the right questions.

Chapters Six through Eight will cover three hydroponic practices used to grow feed for livestock - microfodder, greenwater culture of duckweed and sprouting. We will cover the general process for each practice and meet some farmers. You will also get enough information to decide if a certain system is appropriate.

This book is presented from an objective viewpoint, except where stated. Yet, there are inherent biases that are hard to ignore. As a reader, it is only fair to share some of them before you choose to read more.

First, I've attempted to strip down the fancy words because they can be a barrier to open and honest communication between people of different backgrounds. However, some words are essential to use so they maintain significance. The term 'significantly' is used frequently throughout this book. It is a word used specifically and intentionally, not to make your mouth full for no reason. In science and other data-driven fields, significantly refer to the differences between data correlated with statistics. Most research holds to the standard that something is statistically significant is when there is a ninety-five percent chance that the data is different than the other data.

Next, my father is a retired United States Department of Agriculture (USDA) veterinarian and a Colorado State (CSU) graduate. He worked with ruminants like cattle, sheep, goats, and bison within the commercial slaughter industry. As a regulator and scientist, dad surfed the slaughter line and animal areas to ensure food safety and animal welfare. He would return from work with boots covered in cow blood. A smell I can tell from a mile away. Or maybe that's the giant slaughterhouse just down the street from where I live?

But this is cattle country and I do have an affinity for ruminants, even though I grow plants with fish and raise egg-laying chickens. I also work alongside Colorado State Extension and value what stems from that school. So one can imagine that an agricultural education in Hawai'i is unexpected.

My passions allow me to use a unique agricultural experience. I was born into a prolific agricultural area. Yet, I learned to farm on an island in the middle of the Pacific. I worked as an extension agent under a Japanese-Hawaiian fish researcher in Hawai'i. A fish researcher that trained me to be a scientist by studying fish fodder. That scientific mindset led me to learn from the oldest research university in the US, so I could better think about the world as a system.

Thinking Outside the Soil is only a start. The start of a conversation with farmers, policymakers, and those involved in agriculture. There are still many unknowns. And then there are the things we don't know we don't know. I believe that farmers must come together to improve certain areas of life.

We Need More Chickens

St. Patrick's Day had always symbolized luck, but for me, this one was special. I stood beside Kat Caswell on a March 17th morning that vibrated with a sense of purpose. It felt like more than luck to be here. Some could have said it was fate, though fate is what we make. Kat, the soil-loving agronomist from above, had joined my hydroponic fodder fan club, partly thanks to a WSARE grant she had secured. So here we were, about to share the stage, teaching a course on "Hydroponic Fodder for Chickens." funded by the USDA.

Her and I came from different backgrounds and philosophies but were united by a common goal. Kat's newfound support for hydroponics symbolized a sea change, not just for her but for everyone willing to entertain a new way of thinking.

I couldn't help but think back to a young man named Kewani and a simple idea that rippled into something much bigger. A young man much like my former self - a wild card society had washed away. He came to me not by a judge, not by a probation officer, but by a teacher who still believed in him.

It was the first Greeley Winter Farmers Market after the 2022 bird flu epidemic had left the nation reeling. Eggs had become precious commodities and Kewani and I sold out almost immediately. That's when, with the clarity that often comes from a kid, he said, "We need more chickens."

"We can't," I replied, pointing out the city laws that wouldn't allow us to have more than two chickens per acre.

Kewani persisted with his chicken math, "Yeah, but Sean, if we had just two more chickens, we could get two more eggs a day and bring more eggs to the next market. That means we could make more money."

The simplicity and logic of his statement struck me. This kid was failing high school math and biology until he met me. So what did I do? I got online and found a hundred mature hens.

A few days later, I drove back from Torrington, Wyoming, with a flock of new mature hens. Kewani came in for his next scheduled shift, looked at me and the new hens, and said, "Sean, I was joking." I insisted, "No, you weren't. And it was a brilliant business plan."

With a cheesy grin, Kewani slipped on his chicken boots, and we were off. Over the next month, we built fences and expanded coops. We took his simple idea and transformed it into a small-scale egg production business on the edge of Greeley.

So, as you dive deeper into this book, consider this: Your St. Patty's Day moment of luck may not be luck at all. It may be a simple idea, waiting for you to act. You also have the power to transform a simple thought into groundbreaking action - A lot like Kat took a chance on hydroponics and Kewani's chicken math,

Don't wait for your four-leaf clover to find you. Be the farmer who plants and waters the seeds for them to grow. Sometimes, the most revolutionary changes come from the simplest ideas.

Drought

Water, water, every where, nor any drop to drink.
- Samuel Taylor Coleridge

The blue planet. A place covered in water. The great physicist Carl Sagan called us a "pale blue dot" in the cosmos. A place unlike any other and "the only home we've ever known." Our place in life relies on water. It is our most precious resource.

Unfortunately, only about three percent is drinkable. Humans use most of the drinkable water in the world to grow food. Over a trillion, trillion gallons of freshwater were used in 2021[2]. This makes sense because we need food to live. The relationship between water and food is critical to any country's well-being, economy, and security.

Humans relies on the cooperation of water, plants, and animals to live. Farmers use the water to grow plants. Plants become food and fodder for consumers. Some consumers are the livestock that helps farmers and feeds most of the world. Ultimately, water is essential to growing animal fodder.

Water is usually collected and stored in different ways for future use. In many parts of the world, the stored water is used to grow crops for farms. A country's irrigation strategies allow for sustainable food production throughout the year. Empires fall and rise with the ebb and flow of their water supplies.

Irrigated agriculture is the world's foremost consumer of fresh-water supplies. The World Bank states one fifth of all land used to grow food is under irrigation. This equals 40 percent of the total food produced worldwide. While seemingly efficient, irrigated crops use most of a nation's water. The US Geological Survey says the US used about 120 billion gallons per day for farm water.[3] Similar trends can be seen in other countries.

In arid and semi-arid regions like Australia, the Middle East and the American West, water is scarce. Water scarcity is when there is not enough water to meet all demands, including that needed to grow food and fodder. A better word would be drought. If the United Nation's estimate of ten billion people by 2050 is accurate, the demand for water will only increase with time for agricultural and non-agricultural uses. As will drought.

Many of us in the Western US have heard this word 'drought'. Current US drought conditions are updated online in the US Drought Portal. The government website has been around since 2008 to

> *provide a one-stop shop for data, decision-support products, resources, and information on drought—from drought monitoring and prediction, to planning and preparedness, to applied research. ... By providing a centralized location for reliable, timely, and accessible drought information, Drought.gov offers an essential drought and climate service, improving the nation's capacity to proactively manage drought-related risks and strengthen communities' resilience to drought.*

The Drought Portal has evolved over time to address a growing public demand for actionable, reliable, and shareable information.[4] It seems every year the US has drought-related conditions and asks its people to prepare for water restrictions.

The US has experienced severe shortages in water supply due to repeated droughts. Natural water resources are affected by global climate change. Many regions have less rainfall and higher temperatures causing rangelands to deteriorate. Decreased grassland yields cause negative impacts on grazing lands. Fields used for hay suffer.

Over the past years, we have been flooded with news about farmers who face drought. Drought poses challenges for sustainable field cultivation. In many arid regions of the world, farmers have depleted groundwater supplies and rely on imported water. Sinkholes show where man has sucked out all the water. Water for those farmers must arrive in tankers or totes at much higher costs. The word drought will haunt farmers for years to come. Those living without water may already know the horrors.

Yet drought implies something temporary. In my thirty-seven years, I have yet to see the drought disappear. There is a better term for this, and it's called aridification. That is a fancy way of saying it's dry and will keep getting drier. The next step is a desert. Funny, many people refer to the West as a desert already.

Popular programs like *Last Week Tonight with John Oliver*, *NPR* and *Vice News* have all featured full length stories about drought. The radio show *1A* ran a story about how Colorado producers are farming in a drought. A story that starts in Greeley, Colorado. Colorado Agriculture Commissioner Kate Greenberg and other locals discussed the state's water challenges. I've only had the pleasure to meet Kate Greenberg a few times when she came through Weld County, but I could recall her stance on drought each time she spoke, her stance on drought was clear: traditional methods and cooperative efforts are the way forward. But could there be more to the story?

Like many traditional water policymakers, the radio show cited advances in resource policies, new sprinkler technologies, and, most importantly, farmer cooperation as key factors that have improved water use efficiency over time. [5] Many advocates of hydroponics have heard this rhetoric before. After the show, my mind raced back to my own hydroponic systems, a living testament to water efficiency. Weld County is an area full of farmers, many of whom could benefit from the very techniques Blooming Health Farms had already employed with chickens.

I had a choice to make. I could sit back and simply be a spectator, nodding along with the majority, or I could take a stand, leveraging this moment to introduce a new paradigm—one that could very well rewrite the script on water use in agriculture. This book is a wake-up call for a new stage for agricultural water.

I contend we cannot mend the old model with a band-aid solution. Farmers need practices that improve crop yields with less water. Methods and technologies to improve water use efficiency and productivity are essential. There is popular interest in hydroponic fodder production. Hydroponic fodder production is the practice of growing field crops into livestock feed using less water and space. Farmers growing hydroponically in the places like the American West use over 90 percent less water and space than the same crops grown in the field to get the same yields.

Go West

Colorado sits atop the waters that feed the entire state along with eighteen others and Mexico. Across the country, it's known as the Continental Divide. A place where water falls and freezes upon the land. The runoff is destined for the Pacific Ocean or the Gulf of Mexico. About 80 percent of Colorado's water falls and flows west of the Continental Divide, while 80 percent of the population and the majority of irrigated acres are found east of the Divide. The Platte and Colorado Rivers provide water from California to the Mississippi - the largest watershed in the United States.

Most of the water that falls upon the land sustains life. Many may think of Colorado as solely snow-capped Rocky Mountains, thanks to John Denver. Yet, upon arrival, many learn half of the state is flat. The Great Plains start at the Rocky Mountains and flatten into a prairie paradise.

It is impressive that the flatland farmers can grow food as well as they do. Farmers and ranchers make it the eighth most productive agricultural area in the United States. Weld County is a farming haven on the flatlands of northeastern Colorado. Next to California, Weld is the largest agricultural producer in the country.

Alfalfa, barley, corn, clover, cowpea, and sunflowers wave in the prairie wind far as the eye can see. Most of what gets grown are livestock and the feed they need to make meat, milk, and fiber products. Amber waves of grain sustain the prairie people.

Weld County is not only the largest agricultural producer in the state. Farmers must use over eighty-five percent of Colorado's water to grow food. Not surprising, considering most water grows food.[6]

The farmers here rely on their cleverness to make water work for them. Ditch diggers from the old world helped bring the snow-capped hills onto the prairie. Before those days, the prairie was a paradise. The natural waterways created the range where the buffalo roamed and the deer and antelope played.

The ditch work done in the early 1900s set up Colorado for creating a robust irrigation system. Today, the South Platte and Colorado Rivers provide a lot of the drinkable water from the Pacific to the Mississippi. A well researched account of the fight over water rights and how Colorado created a model irrigation system is in *Confluence: The Story of Greeley Water*[7].

Today, the Colorado from the early days leaves a permanent dry mark on the shorelines of Lakes Mead, Powell and Havasu. It is clear that water is an issue as populations grow in desert cities. Areas with historic aridification where water must get used to grow food for us and animals.

How Important Is Animal Agriculture?

Animals' contribution to the current food production is important to consider. From a global standpoint, livestock products are a priority for promoting global food security in the face of climate change. Animals provide an economic foundation for most farmers.

Animals also represents an important source of capital for farmers. The USDA reports on agricultural statistics for the US. Animals make up about 45 percent of the United States' agriculture.

Globally, it is higher at about 55 percent. Countries that are less developed yoke the burro, bull or horse to work the land. Animals perform many things machines do in developed places. Many provide companionship, and most help create the world we live in. Animals and their products contribute to many things that we use daily and are integral to the livelihoods of humans. This contribution should not be overlooked.

Animal agriculture is a noteworthy contribution to a country's economy. In the United States, animal agriculture employs more than 16 million Americans. According to the US Labor Department, that's about 10 percent of the working population. Almost a quarter of the total agricultural GDP is from related exports.[8] According to the USDA, "Farmers' livestock, dairy, and poultry sales account for over half of US agricultural cash receipts. Since 2015, cash receipts from animal products have exceeded $160 billion per year." Annual US exports of animal products have a value of $31 billion dollars.

The current United States diet includes large proportions of fruits, vegetables, and animal foods. However, more and more consumers are transitioning to plant-based or vegan diets. Environmental issues and the cost of livestock production are major concerns. Some leading authorities suggest that removing animals and animal-based products would reduce pollution and enhance a country's food security.

It is possible to meet an individual's nutrient requirements with plant-based rations. However, A literature review on the environmental impacts of different diets found that plant-based diets that reduce greenhouse gases (GHG) were also often higher in sugar and low in essential micronutrients. Plant-based diets with low greenhouse gases may not improve the nutritional quality.

There are many plant-based diets. But, this can be a challenge to achieve in practice for an entire population. Radical change to an entire agricultural system may be a bit much. However, sometimes science likes to assume the extreme to tease out answers.

Robin White and Mary B. Hall, out of Virginia Tech, decided to answer some of these questions. In 2017, they looked at the nutritional and greenhouse gas impacts of removing animals from US agriculture. The study simulated scenarios based on readily available plants that can be grown in the current system but kept by companion animals and workhorses.

Livestock Impacts on The Environment

Simulated removal of farmed animals could potentially increase the overall food available by over 20 percent. Without animals, there would be a dramatic increase in the grains and legumes. Farmers could then feed those grains and legumes to hungry humans. Tillable land could be converted to produce more food for people.

The study compared different combinations of diets with food currently available. But, when researchers removed animals, supplies of certain nutrients needed for human health were missing. Despite attempts to meet nutrient needs, available foods cannot meet certain requirements. Data from the Center for Disease Control (CDC) also shows that plant-based rations tend to be more deficient in Protein, Calcium, Vitamin A, and Vitamin D.

Several nutrients came up short. Vitamins A, D, E, K, and B12 were deficient in all simulated diets. The studied diets lacked enough calcium, choline, Eicosapentaenoic acid (EPA), Docosa-hexaenoic acid (DHA) & arachidonic acid. The USDA and Food and Agriculture Organization of the United Nations (FAO) recommends omega-3 fatty acids like EPA and DHA for the positive health benefits of lower heart disease and visual acuity. Plant-only systems seemingly produce foods with fewer micronutrients.

A new idea emerges that helps address a big-picture challenge. Micronutrients, rather than macronutrients, are a critical challenge in scaling diets from a person to a population. This idea supports the role of animals as part of agriculture. We need livestock to have diverse nutrition and an abundant ecosystem.

Animals are Part of The Environment

Fortunately, the animals found on farms can be our vegetarian cousins. They can use a plant-based diet differently than we do. That is why the products produced from animals have added value to an agricultural system. Unfortunately, that value is often overlooked by popular pollution narratives.

This may be because almost half of all agricultural emissions in the US are associated with animals. By comparison, for human-used crops, 40 percent of emissions are due to grain production, 5 percent to grow vegetables, and 2 percent to fruits and nuts. Researchers at Virginia Tech showed that the greenhouse gases could be reduced by almost 30 percent by eliminating farmed animals.[9]

Most of the pollution is in the form of methane, a potent greenhouse gas and natural byproduct of how livestock like ruminants process food. An NPR report in early 2022 showed that what cows eat can reduce a big source of GHG emissions. Livestock production produces most of the global greenhouse gas emissions.

Even the United Nations lists animal agriculture as a major contributor to the greenhouse gasses that contribute to global warming. Livestock production represents about a sixth of total human-induced pollution. Methane and manure from ruminants represent a significant source.

Haber Process

The world's agriculture system blossomed after the advent of the Haber-Bosch process. The Haber-Bosch process is a chemical method for making ammonia from gases in the air. It has revolutionized agriculture by providing a reliable source of nitrogen-based fertilizers, which are essential for plant growth. The available synthetic fertilizer allowed anyone to grow plants. Moreover, it could be put on a tractor and applied as a liquid or a dry powder. This made it easy to use and widely adapted by old and beginning farmers alike.

Plant scientists have long known that synthetic fertilizers dictate plant growth. So it is no surprise that two-thirds of the ammonia manufactured goes on farmer's fields. The US consumes 12 percent of the world's annual synthetic nitrogen fertilizer production.

US farms also exploded in the 1930s. Those looking to farm found places like the Central Valley of California or the Great Plains. Fertilizer gave new life to the soil and yield never seen before. Better living through chemistry. The Great Dust Bowl left lessons on how not to farm.

Yet, reliance on fossil fuel inputs to make fertilizer makes food systems vulnerable. The Haber-Bosch method relies on fossil fuels. For example, manufacturing ammonia-fertilizer commonly uses natural gas and gasified coal. Resources that are often criticized for the pollution they produce.

Thanks to crop, soil, and water management improvements, we can better apply fertilizer to the ground. The burgeoning field of precision agriculture is helping farms become the future. The same precision used for hydroponic fodder. And better application means we mine or drill for fewer minerals. It means we potentially pollute less. But many farmers still apply too much fertilizer.

Choking The Gulf

"Sorry I am late," I exclaimed. "I got hit by a train, well, I mean I got stopped by a train..." I trailed off as some of the students giggled. I stood on a stage to present fish farming systems to grow plants. The giggling students were first-year engineering students at the Colorado School of Mines. The Colorado School of Mines is one of the premier engineering schools in the world. They are most known for mining and engineering.

Mines has a program for freshmen called EPICS. EPICS is a design course that has a theme every year. Each school semester, first-year Mines students tackle a different real-world problem. They get taught technical, open-ended problem-solving skills. Students spend part of the program learning from diverse subject matter experts for a capstone project. The theme of that year was 'Food Sustainability and Food Deserts.'

The class was tasked to create a solution that addressed food sustainability. I presented about the importance of food security and how using fish to grow food can address food deserts. I covered the principles and showed the class different scales of food production systems.

I pointed out a challenge to our current agricultural system. A large picture of the Gulf of Mexico appeared during the presentation.

The students heard me say, "In the spring, 7.8 million pounds of nitrogen fertilizer flow down the Mississippi river every day. As a result, intended nutrients turn into pollutants. Disastrous effects occur when they accumulate in the Gulf waters. Beaches and Fisheries shut down due to pollution.

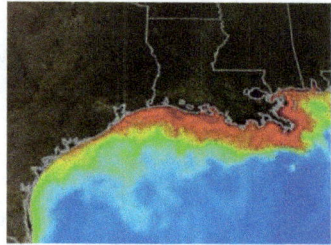

1. Satellite view of the Gulf. The darker area near shore represents areas where a nitrogen-fed algae bloom has risen.

People in the Delta see algae blooms that choke out life in the Gulf of Mexico. The runoff promotes algae blooms and leads to mass fish kills that destroy the ecosystem. Decreases in the fish and seafood catch require more imports of these products. "I never thought to ask, "What happens if the boat doesn't come?"

I ended the talk with a challenge to get out into the field and learn from real-world users. Those in the field are essential to finding real-world solutions. Using novel ways to grow food or fodder can address food deserts. Water scarcity and pollution challenge our food systems. We must look to farmers for different solutions to help solve some of our agricultural issues.

One Solution is...Hydroponics

Meeting the demands for animal products with conventional practices is reaching its limits. Innovative approaches with new technology applications are imperative to address climate challenges. Targeting the feed chain of animal agriculture could help tackle the challenges farmers face. Hydroponic fodder production is a solution that saves water and land and can reduce pollution.

It is important to understand the potential of hydroponic fodders' value to agriculture. These technologies can address environmental issues like drought and pollution. This book's holistic presentation may help identify the best policies that optimize future benefits.

A valuable tool for understanding new or unknown technologies is the case study. Case studies look at particular situations to draw out conclusions and correlations. For example, examination ensures hydroponic fodder systems do not exceed any benefits. Fortunately, hydroponic fodder production is already in use.

In 2021, Canadian researchers examined hydroponics' potential for reducing pollution in livestock agriculture. The study compared hydroponically grown barley to field-grown barley grain feed. It focused on barley because it allows for a straightforward analysis based on other data.

The research took place on an indoor farm in Alberta. Fodder grew in an automated vertical pasture hydroponic unit from HydroGreen Global Technologies.[10] Although HydroGreen declined to comment for this book, what follows is a detailed examination based on their work.

The HydroGreen fodder system creates a large volume of feed in a small space. With the push of a button, barley travels from a bin to a growing area. Lights let the fodder turn green while water falls from overhead sprayers for six days. In less than a week, the system produces mats of green matter. Researchers, rather than farmers, studied the slabs of fodder for analysis instead of for feed.

A single 6-tiered fodder system like the one in the Canadian study can produce over seven-hundred-fifty dry matter pounds of fresh forage daily. Since the system is harvested at one level daily, there is a potential for over two-hundred-seventy tons per year.[11] This is a great deal different than field-grown barley. The Small Grain Report from 2021 cites an average of one and a half tons per year. Soil farmers might get two barley harvests per year.

What's more, scientists confirmed that the system used ninety-five percent less water and space than field-grown barley. Although this makes for an efficient farming system, it requires energy and thus has associated emissions. It is important to recognize that pollution is also associated with hydroponic systems.

Energy data associated with using the fodder system came from the farm. The primary source of pollution comes from the energy needed to operate the hydroponic system. All animal farms need energy to operate. Renewables like solar and hydro may offer answers.

However, the HydroGreen systems show significant reductions in farming inputs compared to grain or hay. The system does not need inputs such as pesticides and fertilizer. Tractors are eliminated, and farm machines are fueled less. And hydroponic fodder farms can lead to less pollution.

Reduced pollution is associated with increased yield in small spaces, a lack of fertilizer input, and no large farm machinery. As a result, there will be a reduction in the potential runoff. There will be less waste that ends up in the rivers and oceans. Pollution and preservation of our natural resources are optimized.

Farmland is spared through more efficient farming systems. The 6-tier system in the study can compete with small-acreage farms and feed at least a hundred cattle daily. HydroGreen's top-of-the-line systems can replace five hundred acres of farmland and feed two thousand dairy cattle each day.

Land Sparing

Hydroponically sprouted barley produces a greater amount of nutrients per area of land. The previous case study showed more than one-tenth of an acre of land is spared for every acre of farmed land that includes a system like HydroGreen's. For example, the 6-tier system in the study can compete with small-acreage farms and feed at least one hundred cattle daily. HydroGreen's top-of-the-line systems can replace 500 acres of farmland and feed 2,000 dairy cattle daily. Hydroponic systems require less land to produce 1 ton of the same crop. Farmland is spared through more efficient farming systems.

Hydroponic operations would not consume extra farmland and does not need extensive land clearing for pasture and feed. Farmers could produce the same amount of nutrients and energy using a part of their area. Hydroponics provides ways for turning cropland into permanent vegetation while getting similar yields. Thus, hydroponics may serve as a strategy for achieving land efficiency objectives. The converted cropland aids carbon capture by enhancing barley production using hydroponic systems, much like how carbon sequestration exists when shifting from tillage to no-tillage. Policy makers say, "what?"

Applying this view to hydroponic farming positions it to save space for other habitats. Farmers can incorporate biodiversity into their farmland. Hydroponic technologies, in a sense, could reduce or limit the expansion of forage and fodder agricultural lands. Imagine less habitat destroyed to feed livestock.

Farmers are frequently left out of climate challenge discussion. Yet, farmers can take the most straightforward steps to help others make decisions. The case study suggests that hydroponic fodder systems are helpful for climate mitigation. Hydroponic systems are the better fodder option for tackling climate change challenges.

Given these compelling advantages, it's time for policymakers to turn their attention to hydroponic systems as a key solution in the arsenal against climate change and land degradation. Regulatory incentives could be designed to encourage farmers to transition to these more sustainable methods. Whether it's through tax breaks, subsidies for hydroponic technology, or more favorable zoning laws, legislative action can play a pivotal role in accelerating the adoption of these systems.

The Grass Is Always Greener On The Other Side

Many people have asked, "How come you left Hawai'i? Isn't that place a paradise?" I reply, "Yea. And, I am from Colorado". Some nod, others dismiss. Yet, it shows me that perspective is key. Countless people peer over the proverbial fence to see the greener grass on the other side.

This can be true for material or mental things. How did their crop grow so well? What have they got that I do not? What do they know? Where can I figure out what I want to do?

Knowledge is power. Many see knowledge as "greener grass" on the other side of what we already know. Knowledge seems to stay around itself. It seems smart people hang out with only smart people. But, someone is always more intelligent than another.

Imagine being surrounded by rocket scientists. They work for defense contractors, research institutions, and think tanks. People who build battleships, send things to space, and consult world leaders to solve crises—the Carl Sagan's of our time.

What if one of those rocket scientists looked at you and said, "Wow, I wish I could farm?

Dr. Judith Dahmann is a reputable professor among her peers at Johns Hopkins. She has developed many international standards and counseled world leaders on many topics. In addition, she works for the MITRE Corporation as the chief scientist in the Department of Defense (DoD) Office of the Under Secretary of Defense for Acquisition, Technology and Logistics. MITRE is a government think tank for the DoD. So she fits the rocket scientist bill.

She asked me to stay after class one day to talk more about what I do. She began with her praise for farmers and expressed shock that more were not looking at farming systems. She commented that she was an outspoken advocate for sustainable agriculture and was part of a land trust. She asked if I was familiar with a land trust.

Land trusts are nonprofit organizations that own and manage land or waters. Land trusts are used for the protection and stewardship of natural areas or for preserving lands for farming. Most land trusts limit commercial development and conserve land, waterways, and nearby forests. Dr. Dahmann's trust also helps to protect working agricultural lands and limits non-agricultural uses of the land.

Onboarding

Do the thing you're afraid to do, and the death of fear is certain.
- Ralph Waldo Emerson

Have you ever started something and felt like an imposter? Have you ever had to "fake it 'til you make it?" My experience getting into Johns Hopkins felt out of place. Never did I think a molecular biologist would study engineering. Especially at a place considered one of the most prestigious in the world. But, my professors and peers showed time and again that they are in awe of farmers.

Perhaps you are just curious. Hours spent on Google or YouTube, wondering about food and where it comes from. You thought to yourself, "I should be growing food this way!" But the questions and doubts form about the ability to farm.

Many begin to farm on a whim. You walk into a feed store and leave with chicks or show up to a friend's farm and leave with a weaning mama and her baby. Maybe your kids want to do 4H and you have a room full of piglets. Next thing you know, you're a farmer. No one would know that you're not.

As a farmer, you are unique in your ability to jump in and learn. That is a trait most farmers and ranchers own. The famous naturalist said, "do the thing you're afraid to do, and the death of fear is certain." Hydroponics may be new to you. That's ok. Because growing hydroponic fodder will draw upon skills you already know.

Most of the time, we don't know what we don't know. When approaching something new, this is usually true. That rings true for fodder systems, even with seasoned farmers. Without fail, the reasons why people abandon hydroponic production is due to lack of knowledge and training.

Like any system, there needs to be a process and procedure. I have learned so far that most hydroponic operations fail because there is a lack of proper training. Training is more than learning how to operate a new system. It also requires us to think differently and form new habits.

When beginning any new fodder operation, it's best to establish a written routine. While many have a stellar memory, the act of writing helps remember the process and allows for a reference later on. A good system should have a set of instructions to guide the user to success.

Another aspect of training that often gets overlooked is teaching others that need to grow fodder. Most farmers teach through action. The trainee shadows and does the work to learn the job. But what happens if that person happens to be gone? What if someone who memorized the process never returns to your farm? The reference is important to ensure successful future operations. There must be proper training. And it will be easy to form habits for success.

What Sets Me Apart

"The more we seek the more kids I get to help"

Sean Short

A consultant or engineer delivers, and then the user is left hanging. Many companies can bridge that gap with excellent customer service. However, something must be developed with you as a farmer. Every system needs time to settle into its unique environment. There will always be subtleties that don't follow suit and deviate from the set instructions.

I take the time to assess your needs and set you up for as much success as possible. That could mean writing a book to explain a new topic. Or it may mean tweaking an established process you're using, but that may not work. Process improvement is important. Especially when time is money. And time is very precious while running or working on the farm.

I approach all challenges with a needs analysis. This generally begins with some sort of operational deficiency. Water and land are scarce. Money is tight. It may also begin with some technological opportunity. Hydroponic systems save resources. Systems increase your bottom line.

This is best captured on an 'Initial Project Needs Assessment.' The assessment gathers basic information about the purpose and scope of a project. It is short and allows for a quick consult without wasting anyone's time. This is where we would talk about if we are a good fit and the next steps. Ideally, I want to advise you through at least one to two crop cycles.

I try to ensure people have the best information to make informed decisions. Years ago, I worked for an aquaponics company called The Aquaponic Source. I did Systems Sales, Design, and Product Support. Part of that job was to tell customers accurate information so they could make a sound investment.

I often talked people out of purchasing aquaponics systems until they got a better experience. Many people wanted to pursue aquaponics because it was new. But, they were unwilling to take the time to learn how to raise fish in a manner that allows good plant growth. I recommended simple systems so people could learn hydroponic principles without a considerable investment. Learning is expensive, and we must use our time and resources wisely.

A hydroponic project can differ in the time needed for completion. Some systems are "off-the-shelf," and minimal time is required for planning. Setup can be done in a few days or less. Extensive projects may take weeks or months to complete and need proper planning. These projects need systems engineering consultation to deliver a workable concept. Design and implementation of a solution. What sets me apart is my commitment to a process that is both helpful and useful. I would love to hear from anyone who wants to pursue a hydroponic endeavor.

In the chapters that lie ahead, we'll pull back the curtain on many of the 'whys' and 'hows' of hydroponic fodder farming. So, when you're ready to redefine what agriculture means to the world, go ahead and turn the page.

"It works if you work it."

Anonymous

---

1. StoryMaps, 2022

2. UN. "Water Facts: UN-Water."

3. "Total Water Use...", 2022

4. "Drought.", 2022

5. 1A. "Water Week: How to Farm in a Drought."

6. Colorado State University. "Water Uses: Colorado Water Knowledge: Colorado State University."

7. Hobbs et al., 2020

8. "Animal Products.", 2022

9. White and Hall. "Nutritional and greenhouse gas impacts of removing animals from US agriculture."

10. Newell, 2021

11. HydroGreen, 2022

Chapter Two

HYDROPONICS

[Hydroponics] has vast potential to provide food in areas of non-arable lands, such as deserts and the high prairies of the American West. Overall, the main advantages of hydroponics are more efficient plant nutrition regulation, availability in regions having non-arable land, efficient use of water, ease and low cost of sterile media, and higher planting density...leading to increased yields per acre.

Howard M. Resh, PhD

The art and science of growing plants without soil.

James Sholto Douglas

Becoming A Hydroponics Guy

The world forever changed on September 11, 2001. I was a high school student and didn't know what I wanted to do with my life. Patriotism took over my young mind and I thought the path was to fight for my countrymen. Much to the dismay of my West Point grandfather, I enlisted in the Marine Corps in the fall of 2003. My recruiter convinced me to pursue Force Reconnaissance once I got in as a Radio Field Operator. The prospect of signing my life away to Uncle Sam delayed my entry for one last hoorah. Instead, I went off to the snow capped mountains to work at a ski resort. Though, I never got to say "oorah."

After a few months in Breckenridge as a ski bum, I fell off of a 30 foot building and broke my back.

Doctors told me I would never walk again. I tried to rehabilitate with the help of physical therapy and a heavy dose of depressants. Vicodin and alcohol kept the pain away. Thankfully I found an alternative in medicinal cannabis.

An overwhelming amount of research and anecdotal information advocated cannabis for multiple medical means. The most notable was the anti-spasm and pain-alleviating properties, which were my most concern. After my medicinal introduction to marijuana, I was able to curb and eventually cease my pain pill-popping problem. When the fog of opiates and alcohol was lifted, I was able to find the motivation to take care of my physical health.

I later lived with my brother in a Colorado mountain town and worked at Breckenridge Ski Resort. He was a mechanic and I on mountain portrait photographer. Even though we lived in a paradise, the jobs paid meager wages to make a living. At the time, our town had the most cannabis shops per capita. I wanted to save money by growing the medicine that helped me get back on my feet.

First, my brother and I went to the local grow shop and spoke to the owner. He asked us questions about what we were getting into, and we gave a common response. "We're gonna grow some tomatoes and peppers," we both snickered. The owner looked at us funny but immediately knew what we were trying to do.

Next, my brother and I scoured the small store. Shelfs were packed with bottles of nutrients and parts to build hydroponic systems. Components like lights, pumps, tubing, and plastic trays of all different sizes.

Finally, we selected a system called 'Bato Buckets.' All we had to do was add this one nutrient, and an air pump would automate most of our operation. We spent most of our money on artificial lighting, and we were on our way. My brother tasked me to be the cannabis grower.

Early Retirement

Well, three months later, the plant had long branches that were sparsely populated with green leaves. The flowers, or buds as they're known, were nonexistent. If one wanted to use it for medicine, there was little use. My pot plant was pretty pathetic.

One wouldn't know from the pictures we took at the time. The smile showed a passion for plants. The cheesy grin suggested I was proud of what I did. But, even though I was happy with what I grew, I knew it could be better.

I immediately dove into the literature to find answers on how to make better plants. I spent countless hours on the internet and in books looking at what others had done. I aspired to be the next cannabis entrepreneur like the pot shops in my town. But I needed to learn from an expert on the subject.

I first found that in a book. The book that opened my mind to plant science was called *Marijuana Botany* by Robert Connell Clarke. Clarke begins with plant life cycles and propagation techniques. Then, he hits me with chapter three: *Genetics and Breeding*. I learned the basics of breeding and how to make more plants. My mind contemplated a plastic view of the plant world. Little would I suspect that a tomato would forever change my life!

One day in my studies, I saw an image that looked like a green tree growing indoors. I was intrigued because it looked, unlike anything I had seen in the plant world. Green, but out of place. The plant branched out in every direction with green woven along a metal trellis. Dangling from the tangled framework were clusters of small, ruby red dots about three or four feet above the floor.

The caption below the picture told me all I needed to know. I was looking at a picture of a plant growing over an entire area. The plant has been recognized as a Guinness World Record Holder, with a harvest of more than 32,000 tomatoes. I thought, "we should be growing our food this way!"

2. Tomato. Epcot Center in Orlando, Fl.

Cannabis influenced me to retire as a photographer and pursue a career in cultivating tomatoes. Retirees have to start somewhere, so it made sense to return to my roots and be in Weld County. I enrolled in an Associate of Science at Aims Community College and grew tomatoes on the flatlands where I was from. I sensed hydroponics was an untapped potential and aspired to start a hydroponic tomato farm.

I built the first hydroponic systems in the backyard of my father's house in Greeley, Colorado. The first system was made from a Rubbermaid tote with netted pots. I then went vertical with cinder blocks and PVC piping. Netted pots grew hydroponic cherry tomatoes and the PVC pipes were ported with holes to grow lettuce and tomato starts.

Inside Job?

The tomato at the Epcot Center and history's first hydroponicists inspired me to farm. My early systems showed me I could grow in unusual spaces. Yet, Colorado has seasons. So if I wanted to be a hydroponic farmer, I would have to make some decisions. Grow when the weather cooperates or figure out a way to make my systems produce all year round.

Initial investigations suggested I should look into greenhouse farming. Greenhouses are simply structures covered by transparent materials such as glass, shade, screen, or plastic sheets. Structures are a common agricultural practice and are popular because they allow farmers to control growth when the weather changes. Today, more than 8,750 greenhouse vegetable farms are present in the US, valued at $9.2 billion.

The best greenhouse would be one suited for my region. High-tunnels are simple structures made from clear plastic sheets and some type of pipe. They are relatively inexpensive and can be temporary. Colorado farmers use high-tunnels greenhouses to extend the spring and winter seasons.

I began by looking at Universities in the Rocky Mountain region that studied high tunnels designed to protect crops from minus 0 F temperatures and potentially heavy snow loads. Environmental factors I had to consider should I want to grow tomatoes.

Isaac Newton reminds us we can "stand on the shoulders of giants" if we want to make progress on our chosen journeys. One of those giants was a Horticulturist at Colorado State University who was well-versed in greenhouses. She gave a presentation I'm glad I did not miss. Every word I heard from the Johnson & Wales auditorium seat about high tunnels or greenhouses made its way from my pen to paper. Afterward, she introduced another giant doing something neat in an abandoned Denver greenhouse.

Colorado Aquaponics

I met JD Sawyer when he was in his mid-30s. He began with gratitude for the presentation venue. Johnson & Wales once employed him as a project manager but had just lost his job in the financial crash of 2008. So he and his wife began to grow his family's food to save money. Like me, JD fell in love with the possibility of hydroponic cultivation.

JD showed that one of the challenges with hydroponics is that the nutrients can only be used so much. You must throw away the "salty" water when the plants use all the nutrients. The brine water would have to go down the drain and end up in the oceans. His wife had heard about a better way. People could use fish water to grow plants. I thought, "fish?"

After JD's presentation, I climbed down the auditorium stairs to make my introductions. His warm smile and passion for his newfound endeavor came through his every word. I saw myself in JD and offered to volunteer at the abandoned greenhouse in Central Denver.

For weeks I drove from my Greeley home to a deserted area called Elyria-Swansea to learn how fish grew plants. We built systems, and I listened to others and their experiences. I learned from subject matter experts like JD, his wife Tawnya, and those at the Grow Haus. And, of course, I read books about hydroponics. I needed to know the science behind growing plants.

History's First Hydroponicists

Hydroponics is a combination of the two Greek words 'hydro' and 'ponos.' 'Hydro' means water and 'Ponos' means to work. So together, they mean 'to use water for work.' However, hydroponics has been around for thousands of years, long before the Greek words.

3. Artist depiction of 'The Hanging Gardens of Babylon

The earliest example is that of the Gardens of Babylon. Detailed descriptions of the Gardens come from the ancient authors Strabo and Philo of Byzantium. These mythical gardens were extravagant. Plants were "*permanently green*" and

> "*the roots of the trees [were] embedded in an upper terrace rather than in the earth. The whole mass supported on stone columns. Waters irrigate[d] the whole garden, saturating the roots of plants and keeping the whole area moist. Streams of water emerging from elevated sources flow down sloping channels.*"

Drought was clearly not a concern for those living in Babylon.

The Egyptians manipulated the Nile river to grow dates and other crops. They relied on the surface waters to provide for the plants.

Figure 3. suggests that Egyptians carried water from some distance and physically watered their plants, which is a good indication that the Egyptians practiced hydroponics. Hieroglyphic evidence of early farmers who used water to grow date palms in sand culture. Not quite the

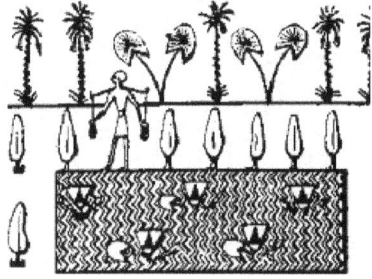

4. Egyptian Hieroglyph of Hydroponic Date Cultivation

shocker that the pyramid builders knew space-age science.

A well-known water engineer was a Greek named Herod. He studied water and used its principle to make water work for society. For example, he discovered how to apply siphons and pressure to power doors and city gates. The water flowing for agriculture could also be used to do other work. Thanks to Herod, we continue to use water for many things, like operating irrigation gates and spinning hydroelectric turbines.

David Sedlak's *Water 4.0* describes the rise of modern water systems and how the Romans built ways to transport water across their agricultural empire. That industrious civilization used water (and war) to conquer most of Europe. Roman engineers moved water through pipes built from lead and stone mostly known as aquaducts. Aqueducts are one of the most well-known ancient irrigation systems in the world. A neat piece of water history is that we get the word 'plumb' from the Latin name for lead, *plumbum*.

The Romans also made areas to collect and store water, which we still model today with pipelines and reservoirs. Colorado ditch diggers got it from someone, right? The ancients even constructed areas to culture aquatic plants and animals. Places across the Mediterranean Sea reveal seaweed ponds and shellfish managed by Roman citizens. People who grew up by the sea knew that animals and water were integral to agriculture.

Close to Home

All advanced societies create clever agricultural systems to be successful. The Aztecs were no exception. The Aztecs had an empire that prospered in part because of their relationship with food. They likely pioneered what is now called urban agriculture.

5. Aztec Chinampas model by Te Mahi, Photographer: Te Papa

Their capital city was situated on an island surrounded by a lake. As a result, the land had to be used with great care. Space for materials, buildings, and people was at a premium. We could say the city had little arable land, yet the city thrived. They had to think outside the soil if they wanted to live. The people used the lake for navigation and sustenance. They relied on the waters to sustain their way of life. Farmers figured out how to grow food on floating gardens called chinampas which used hydroponic principles.

Chinampas were "a raised field on a small artificial island on a freshwater lake surrounded by canals and ditches. Farmers [used] local vegetation and mud to construct chinampas."[1] They floated these gardens around the city and grew things like lettuce, tomatoes, corn, and cabbage on waterways that surrounded their cities.

Water wicked through the floating gardens to feed crops-reducing the need for extra irrigation. Chinampas were effectively individual ecosystems that optimized greenhouse gas emissions and promoted biodiversity. Today in modern Mexico ancient chinampa areas are commonly used for cattle feed.

Chinampas serve as a model for sustainable field production and hydroponic fodder. There are thousands of lakes across the US that have surface water available. What if communities embraced using reservoirs as food systems rather than recreational playgrounds? Why not both? I challenge policymakers, municipalities and farmers to speak up for a chinampa resurgence.

Genie in a Bottle

The Science of Hydroponics evolved in the 1600s. The renaissance gave rise to the thinker minds that likely wanted to farm better. Scientists of the time surely wondered, "How did our ancestors make plants grow?"

Some of the first inquiries to answer those botanical mysteries come from the Belgian Jean Baptiste van Helmont. He took a willow tree, planted it in a pot of measured soil, and watered the plant over five years. Then, half a decade later, he reweighed the soil with startling results. He found very little soil loss, and he was pretty perplexed. Baptiste van Helmont believed that water gave all the things a plant needed to grow.

Scientists soon showed that water dissolved the minerals in the soil. They concluded that plants use a handful of elements and CO_2 from the air. Therefore, they do not need dirt to grow. Yet they were unsure of other elements plants could contain.

Later in 1699, the American John Woodward studied plants grown in glass bottles. The goal was to figure out what exactly plants used. That research helped German scientists Julius Von Sachs and William Knop uncover components of plant life. Before long, they came up with a list of elements. In 1859, Knop was credited as the first to use a chemical solution to grow plants.

Imagine starting your hydroponics journey with something as simple as a jar and a handful of seeds. A method so straightforward yet so potent, it can feed livestock with nutritional benefits that will astonish you. Intriguing, right?

We'll explore this—and how it's transforming lives beyond the farm—in an upcoming chapter. You'll even discover a simple way that streamlines the hydroponics process. But first, I need to share how I came across the Extension scientist that sprouted my agricultural education.

My Desire Farm Fish

I first planned to attend Colorado State University with the ambition to follow in those Extension Agent's footsteps. However, a Google search with the keywords 'aquaponics' and 'Hawaii' led me down a path I never expected. The words 'Small-Scale Aquaponics', 'Harry Ako,' 'CTAHR,' and 'Pacific Islands;' - popped up on the results page. So I clicked on each link and opened them in their own browser pages. Isn't Google great?

Established in 1907, the College of Tropical Agriculture and Human Resources (CTAHR) is the founding college of the University of Hawai'i at Mānoa in Honolulu, Hawai'i. Located in the Mānoa Valley on O'ahu, the College focuses on tropical agriculture, food science and human nutrition, textiles and clothing, and human resources. It's work precedes the creation of the university's flagship campus. I discovered that the Department of Molecular Biosciences and Bioengineering developed sustainable hydroponic models for farmers in American Sāmoa, Hawai'i, and beyond.

The articles showed that an Extension Agent named Dr. Ako designed a cropping system based on recent research out of the Virgin Islands. Ako is an accomplished aquaculturist that developed feeding protocols for the aquaculture industry. Early work by Dr. Ako and a graduate student figured out how to operate a hydroponic system using simple methods.

First, Dr. Ako analyzed a farm system by looking at the mass balance of the system. Then, he used an equation to operate an aquaponics system called the Nutrient Flux Hypothesis. Approaching an aquaponics system from a mass-balance perspective planted the seed for me to become a Systems Engineer. His research helped Pacific Islanders create systems that could be 100% sustainable and not rely on cargo ships.

What If the Boat Doesn't Come?

My first foray into animal feed forced me to become a better scientist. One of the challenges we were trying to overcome was the idea of "what if the boat doesn't come?" The challenge with being on an isolated island is that they rely primarily on external resources for survival. Predictions suggest two weeks of resources if the cargo ships ceased docking.

I had the opportunity to help with research on how to make fish food. Many Sāmoan farmers fed pig slop to their fish and saw terrible plant growth. Fish would also die or would grow very slowly. Research revealed a Starkist tuna factory nearby, and Dr. Ako formulated a homemade diet from the fish meal waste. The work we did took this formulation and developed a feed that was on par with commercial feed. Dr. Ako trained me to help farmers source materials locally.

The years 2020 and 2021 showed the world the fragility of the global supply chain. Within months of a shutdown, grocery stores were empty, and people began to go hungry. Cargo ships sat at sea, and trains stayed in the station. If you didn't grow your own food, you ate what was available. That could have been canned and processed food. Food that survived the supply chain fiasco.

Many in the public eye failed to see, the same was true for animal feed. Agricultural supplies delayed crop production cycles. The transportation of feedstuffs slowed and halted. Bags of feed were nowhere to be found at the feed store. Challenges that reveal our food chain is broken and must be mended.

Thankfully, farmers have proven to be most resourceful people in the world. But what can you as a farmers do? With a little guidance, you can help revolutionize agriculture as the Bengal farmers did with Sholto's Science of Hydroponics.

CHOPKNS CaFe

Plant research since 1859 reveals there are 21 elements that plants need to grow efficiently as possible. Three of the elements are considered nonmineral elements. These are hydrogen, carbon, and oxygen. These elements are abundant in the environment and from the atmosphere. So plant physiologists don't consider those to be part of the essential elements. They get hydrogen, carbon, and oxygen from the atmosphere and soil environment.

Then there are minerals that we call the essential elements — minerals are essential for plants to grow. There are also four other elements that scientists have found that are beneficial for plants and help plants grow better. This information may be helpful for new or beginning farmers with field crops. Plant Scientists today call them the Essential and Beneficial Elements in Higher plants.

There are several letters on this page, and it can be quite confusing to figure out what plants need. While studying at the University of Hawai'i, I had a silly professor that threw a jumbled set of letters on a PowerPoint presentation. Then, he told us was that we had to head down to the Sea Hopkins Cafe. We were to meet Cousin Bomo the Mighty Man and his girlfriend Sakona. Apparently they knew the best way to learn elements. Right near the beach.

CHOPKNS CaFe CuSeZn BoMo MgMn SeCoNA

So right after class, what did we do? We asked to find the Sea Hopkins Cafe. We should have realized the joke when the directions came with a chuckle—what a clever way to use mnemonic devices for plant elements. Silly things allow us to remember some of the more important aspects of stuff we're trying to study. Sometimes it's serendipity that creates memories. Like being born on the same day or attending the same university of the Father of Hydroponics.

Hydroponics is Born

Hydroponics was born over eighty years ago in 1937. A TIME article from March 1st of that year begins, "Last week a new science was given a new name. Hydroponics, by its foremost U. S. practitioner, Dr. William Frederick Gericke of the University of California." And like that, an Extension Agent was credited for coining the term 'hydroponics.'

Coincidentally, Gericke attended Johns Hopkins before Berkeley. He also wanted to tell others about hydroponic cultivation. His first published paper was "On the physiological balance in nutrient solutions for plant cultures" in the American Journal of Botany. In 1929, he published the 200-word article "Aquaculture: A means of Crop-production." During his time in California, he devised methods based on his knowledge of agriculture and plant science.

Gericke took the essential plant elements, dissolved them in water, and then suspended the plant roots in that solution. TIME magazine tells us "netting [was] stretched over the top of the tanks and packed with excelsior or sawdust." Gericke knew that hydroponics used less water because the plants always had access to water.

Hydroponic plants reuse the water and transpire the rest. The roots have greater access to water and can grow closer together. You can plant more densely because plants do not have to draw their nutrients from the soil. As a result, plants use less space compared to soil agriculture.

The March 1st report went on to read, "the seeds are planted and from which roots sprout down into the water. This bed of litter on the netting serves to support the stalks after

6. W.F. Gericke and his wife with a 12 ft. tall hydroponic tomato.

the plants are grown. He got phenomenal results. Each tank has an area of .01 acre. In one of these Dr. Gericke grew 1,224 lb. of tomatoes...Tomatoes are Dr. Gericke's joy." Despite the fame from TIME, it was an unknown Extension Agent from across the Atlantic that helped hydroponics grow, like Gericke's tomato.

Open-Source Hydroponics

By the 1940s, demand for hydroponic cultivation was worldwide. The 1949 ed. of 'Hydroponics: The Bengal System begins with ""[H]ydroponics recently assumed such large dimensions that a popular, practical, and non-technical text dealing with the method has been urgently called for."

The author is probably the most un-known person to have propelled the hydroponics industry. But the rise of commercial hydroponics can be at-tributed to an Extension Agent from Oxford University named Dr. James Sholto Douglas. He stood on the shoulders of scientific giants to bring practical knowledge to those in the field.

7. J.S. Douglas holds crisp and firm ridge cucumbers grown in London

Farmers and Dr. Douglas did years of work to build upon the principles of Dr. Gericke and the military. Sholto showed a repeatable process to grow plants without soil consistently. Work began by figuring out what farmers need to be successful. Then, Douglas designed systems that suited their individual needs and helped them get started. He stayed to tweak the process until it worked and went on with what he knew to tackle the next challenge. This agile approach gave the flexibility to find success.

The Bengal System was born as one of the first former training manuals. The book outlines background information that is need-ed for farmers to be successful. Then the author applies lessons learned to give the farmer tools to grow plants without soil. It con-cludes with calls for more work to be done and provides hope for the future of hydroponic production. Douglas uses new knowl-edge from farmer research to create a science education. Soon after, the commercial hydroponic industry in India and elsewhere erupted.

8. Commercial Hydroponics Farm in Bengal India

Be All You Can Be

World militaries like the US Army also noted the success of Dr. Gericke. The United States was at war and needed to feed its remote troops during this time. War meant the supply ships were at risk from an ocean enemy. It was not practical to ship fresh vegetables while in open conflict.

Hydroponics was a logical choice for an island environment. Most occupied islands were small, rocky outcroppings void of complex ecosystems. As Islanders know, freshwater and space are precious commodities. So the Army adopted proven cultivation practices while at war in the Pacific Islands.

9. 1952 Newspaper article.

The United States built hydroponic units at island bases. The rocky islands could not support soil-based food production. Military bases adopted hydroponic units to grow food for stationed troops. Systems sprung up on O'ahu, Iwo Jima, Okinawa, Honshu, Wake and Ascension Islands.

The Army even created a special Hydroponics Branch. The Headquarters in Japan boasted an 80-acre area devoted to vegetables, one of the world's largest hydroponic farms.

Troops ate things like tomatoes, peppers, and salad greens. Troops were tasked to be the stewards of fellow soldiers. In 1952 they grew over 8,000,000 lbs. of fresh produce. It's a shame they disbanded that division. US Soldiers transformed desert islands into rich fertile sites in mere months. Imagine what they could do today?

10. US Soldier inspects Tomato Beds on Ascension Island

Same Thing Expecting Different Results

Countless sources claim the benefits of hydroponics and tell of a future in agriculture. And while I agree, many claims came before my time. Yet here we are, still using a 12,000-year-old model to grow food. We're seemingly doing the same things and expecting different results.

Reducing on-farm water use and the economic productivity of the agricultural sector is a challenge. The success of farmers and the advent of plastics gives hydroponic agriculture a considerable step forward. Concrete and metal piping could be replaced with plastic, further reducing capital and operational costs.

Hydroponics is a well-known technique for low water use, high yields, and year-round production. Companies like HydroGreen have already implemented systems for farmers. These systems are growing large amounts of barley, among other grains and legumes. Farmers control the prices and quality of feed and see more consistent results when going to market.

Work in California, India and the Pacific Islands showed that a new practice could be effective and practical. It is for this reason Douglas called hydroponics ""the art and science of growing plants without soil." We must follow in the footsteps of Gericke and Douglas to put in place simple and effective hydroponic practices-like growing lentils in a Powerade bottle in my Pacific Y.M .C.A. dorm. That delicious experiment led me to ponder: What other simple, yet impactful, methods are out there waiting to be discovered?

---

1. Ebel, 2020

Chapter Three

WHAT THE F*DDER!?

"Understand where your food comes from and make an informed decision before purchasing"

Josh Ciardullo

The Pivotal Call: An Unexpected Revelation

It seemed like an ordinary July afternoon in 2022 while the sun shined through my office window. Then the phone rang, breaking my silence. The number was unknown, but intuition urged me to swipe and answer the call. On the other end, a gentleman yelled into the phone.

"Is this Sean?" "Yes this is him. How may I..."

"My name is Josh", he interrupted, "and I just love what your trying to do!"

He explained that a local agricultural agent forwarded my contact after he heard about me. But Josh didn't wait to mince words or make small talk. He told me his cattle were on the brink of a crisis. Drought, that ever-growing evil of the modern farmer, was strangling his operation. His cattle were thirsty, and Josh was at a crossroads. Cull or figure it out.

This cowboy's passion for sustainable agriculture was as infectious as mine for hydroponics. Before we set up a time to meet, Josh shared his vision of being a better rancher. His call wasn't just a plea for help; it was a challenge for change in our industries.

Ciardullo Ranch is a 33-acre regenerative operation located in Wellington, Colorado. Josh, his wife Erica, and their young son Henry are not just running a ranch but creating a legacy. Their mission is to work with the land, to respect its ebbs and flows, and to raise their livestock in a setting that mimics the natural world. They've expanded their operation to include 500 acres at Sylvan Dale Ranch, where they practice agriculture like a religion. Ciardullo Ranch focuses on Black Angus, which are grass-fed and pasture-raised. But as I saw during my first visit, no ranch is complete without some chickens, sheep, and livestock dogs.

The conversations with Josh and Erica were a watershed moment. Drought was no longer something the weatherman told us about every day. It was a daily struggle in the lives of farmers and ranchers who feed our great nation. If hydroponics could help his crisis, it would redefine the entire agricultural industry. But I still needed to know, could hydroponics offer a lifeline to other ranchers who raise our Earth's livestock?

Imagine a farm where cows, chickens, and even exotic animals like alpacas abandon their feed for hydroponic fodder. It's not a mere vision; it's a transformative reality happening in farms from Colorado to remote corners of the world. This chapter isn't just a guide. It's a calling for a new era in sustainable farming. A call that can significantly cut your feed costs, improve animal health, and make your farm resilient against climate change.

Many farmers worldwide grow their own fodder. Research, experience, and economics show that hydroponic fodder is an amazing supplement. And a great addition to traditional feeding regimens. This chapter outlines the animals in agriculture that are successful with hydroponic fodder. And where there's room for improvement.

Chicken Eat Anything, Right?

With over a decade of experience in hydroponics, I thought I had uncovered all its benefits. Yet, it wasn't until I started raising my own chickens that I learned how practical hydroponics can be for every farmer.

My first products as a new farmer were organic alfalfa sprouts, organic clover sprouts and organic pea shoot microgreens. Many failures lead to wasted sprouts, so like any resourceful farmer, I fed the spent sprouts to my chickens. Naturally, the chickens ate it up! My chickens didn't just like the sprouts; they thrived on them. They are healthier, and many farmers comment when they see my hens.

12. Rhode Island Red. Photographer: Maxine Novick

I also began to hear my egg customers say, "These are some of the best eggs I've ever had!" I brushed it off at first. But soon understood how Kewani sold out so fast on that fateful Fall day at the Winter Greeley Farmers Market.

As I crunched the numbers during routine farm accounting, the financial benefit became obvious. My feed costs had plummeted by 50%, thanks to the reduced need for commercial kibble and bulk seed prices. This experience was not just a eureka moment but a natural progression in sustainable farming that inspired me to write this book. It also underscored a critical global trend: the rising demand for poultry.

The visual display I first found on StoryMaps showed global chicken consumption is on the rise, and the demand for eggs will keep chickens in charge of number the one spot. Current estimates indicate there are 2 billion chickens in the US and there are over 22 billion chickens in the world. By comparison, cattle come in at number two with over 1.5 billion. That's three chickens for every person in the world or six for each American. [1]Since chickens are on every continent except Antarctica, poultry farmers must consider the hydroponic fodder avenue for their hens.

POULTRY

Hydroponic fodder can feed chickens, geese, turkeys, ostrich, emu, ducks, and other poultry. If it has feathers, it loves hydroponic fodder. Though birds prefer certain crops, poultry performs well when eating fresh hydroponic fodder.

Precision Agriculture

When we talk about hydroponic fodder for poultry, we're talking about a form of precision agriculture. For instance, chickens need a different ratio of amino acids compared to ducks or turkeys. With hydroponics, you can grow specific varieties of legumes or grains. These are rich in essential amino acids like lysine or methionine, providing a more balanced protein source. Hydroponic systems can be automated to produce crops in cycles that align with the birds' nutritional needs. Automation allows you to harvest and feed when the food is most nutritious.

Of note, ducks are the fourth most raised farm animal in the world. Estimates put their population at around 1.2 billion. Most of the ducks are raised in China and Southeast Asia. They are also raised regionally in certain parts of Africa, Europe, and Russia. Ducks are also often kept on small scales as an integral part of a farm's aquatic ecosystem.[2]

Ducks play a neat role in sustainable agriculture systems, often helping to control algae and insect populations. Hydroponic systems can be integrated with aquaculture to create a closed-loop system. The water from the duck pond is used to irrigate the fodder crops. This not only recycles nutrients but also reduces the overall water footprint of the farm. Particular plants that thrive in such setups could be duckweed, an excellent fodder option for poultry.

At Blooming Health Farms, we've adopted a similar integrated approach with our chickens. We use a closed-loop system, which is commonly called aquaponics. This is where the wastewater from our sprouting process is used to grow duckweed and reduce our total water usage.

The duckweeds that thrive in our system are not only ideal for cleaning the water but also serve as excellent fodder for our chickens. This creates a sustainable cycle that benefits the entire farm ecosystem, demonstrating our commitment to innovation and sustainable agriculture at Blooming Health Farms.

The Golden Ticket to Great Eggs

Eggs are gold for many, especially chicken producers. Hydroponic fodder gives your ladies a thicker shell. This is valuable as it will lead to fewer cracked eggs when collecting.

Let's start with shell thickness. A thicker shell isn't just about reducing breakages; it's also a sign of sufficient calcium and phosphorus in the diet. Farmers feeding hydroponic fodder also found more, larger eggs. [3] However, producers note that layers need enough calcium and grit in their diet. Or the larger, thicker eggs won't form. I throw my sprouts and duckweed on the ground so my hens can peck and scratch. Evidence shows that alfalfa, cowpea, and sunflower shells have high calcium levels.

Gettin' Their Peck On

In the study of animal welfare, there is something called "behavioral enrichment." This means changing an animal's living space to make their lives more exciting and fun. For birds like chickens, one of their favorite things is pecking for food on the ground. When we spread out hydroponic fodder for them to peck at, we're doing more than giving them something to eat. We're also helping them do something they naturally enjoy, which makes them feel good and calm. This not only makes our chickens friendlier with each other, but it also means they're less likely to pick on one another and get stressed out.

Micronutrients

Hydroponic systems can be tailored to produce plants rich in specific micronutrients. For instance, sunflower is particularly high in selenium, a trace element vital for immune function. Similarly, corn is rich in Vitamin A, which is vital for vision and reproductive health. By selectively growing these plants, you're not just feeding your birds; you're nourishing them. Farmers see better egg quality, faster growth rates, and reduced disease susceptibility. Yet, more research is needed on mixed feeding regimens to provide essential micronutrients.

Farmers also see faster weight gain in broilers. Researchers in Iran studied the effects of barley sprouts on broilers. They found that a 50 percent diet finished one to two weeks earlier than chickens fed a conventional or 100 percent fodder diet.[4] Imagine a small-scale operation of 500 broilers. Suppose each bird reaches market weight a week earlier and consumes less feed in the process. In that case, you're looking at significant savings.

For instance, let's say it costs $1 per day to feed each bird, and you save 7 days of feeding, that's $3,500 saved in just one cycle. Multiply this by 4 or 5 cycles per year, and you're looking at a significant economic impact.

Not Just Pretty Birds

A bird's feathers are not just for show; they're functional components that help regulate temperature and enable flight. The glossiness of a feather is a result of its structure and the oils that coat it. The bird's uropygial gland produces these oils, and its diet directly influences their quality. Omega-3 fatty acids are found in hydroponically grown sunflower, leading to glossier feathers.

Overall, birds have a better appearance and improved health. Their feathers are shiny and fall out less when stressed. They run to get at any fodder, much like the quoted goats at the beginning of the chapter. My chickens will abandon everything for alfalfa, barley, clover, duckweed, and sunflower.

Each of these details contributes to the intricate tapestry of sustainable poultry farming. Next, Let's dive deeper into the multifaceted benefits of hydroponic fodder for horses.

HORSES

Hydroponic fodder is a powerhouse of balanced nutrition. It provides macronutrients like protein and a wide array of vitamins and minerals. The biological process of sprouting unlocks nutrients in the seed, enhancing their bioavailability.

The high moisture content in hydroponic fodder plays a role in maintaining hydration, making it different from grain and dry feeds. For endurance horses, where water balance can significantly affect performance, this is crucial. The high water content aids digestion, helps move food more through the gastrointestinal tract, and reduces the risks of colic. Because fodder can be up to 90 percent water, it significantly improves a horse's hydration. Water is essential for health and vital bodily functions.[5]

Starch Digestibility

Grain and concentrated feed can be very starchy. Starch takes a long time to digest and can make a horse feel full. You may notice that your horse may not nibble on hay all day. Because other feeds take so long to digest before they are broken down in the hindgut. This can lead to serious health problems like colic and laminitis. The process of sprouting hydroponic fodder turns starch into sugar which is much easier for a horse to digest.[6]

No Mo' Colic?

Colic is often considered the nemesis of the equine world. Factors like diet can contribute to its occurrence. Hydroponic fodder helps the intestines by reducing the risk of impactions or spasmodic colic. Moreover, the enzymes help to pre-digest the fodder, reducing the workload on the horse's digestive system. Thereby lowering the risk of colic and gastric ulcers. Owners that feed hydroponic fodder see reduced instances of colic, ulcers, and inflammation. Feeding fodder alleviates many of these causes, and horses can take full advantage of the nutrition.[7]

Mental Health

The mental well-being of a horse is as important as its physical health. Horses are natural foragers, and grazing is a form of mental stimulation. As seen in Image 13, hydroponic fodder is fed out in grazing boxes, allowing the horse to display natural foraging behavior. This can improve temperament, lower stress levels, and better mental well-being.

13. Barley micro-fodder after 7-day growout

Hoof and Coat, and Reproductive Health

A horse's coat and hooves are often the first indicators of its internal health. The biotin and Omega-3 fatty acids in hydroponic fodder can contribute to stronger hooves and a shinier coat. These are not just aesthetic benefits. A strong hoof wall is less prone to cracks and infections, and a healthy coat provides better thermoregulation.

For breeding horses, the regularity and predictability of heat cycles are influenced by nutrition. Hydroponic fodder can contribute to more consistent and earlier heat cycles, aiding breeding programs.

Fueling Performance

It takes less energy to digest fodder, and the sugar content converts to energy easily. Therefore, horses can redirect the point they are not using for digestion, and the added energy from the sugar contributes to increased energy and power. The amino acids and essential fatty acids in fodder also help build muscle tissue on top of the vitamins that contribute to natural healing and a boosted immune system. With the energy gained from a fodder diet, your horse will be able to recover faster. As a result, horses will have a consistently higher energy level on fodder.

> *"We can say with a high degree of credibility that after being fed with our sprouting fodder... the win and place ratio was better than ever before."*
> Brian Rowe, Licensed Racehorse trainer for 25 years

The Competitive Edge

For racehorses, every fraction of a second counts, and their feeding regimen can be the difference between victory and defeat. The high bioavailability of nutrients in hydroponic fodder means quicker absorption and utilization. This provides an immediate energy source for high-intensity activities like racing.

Racehorse trainers report better win-and-place ratios after adding hydroponic fodder to their diet. This is more than anecdotal evidence; it attests to the fodder's impact on racehorse performance.

The Importance of Post-Race Muscle Recovery

Racehorses undergo rigorous training and intense physical activity, which can cause muscle fatigue and strain. The high-quality proteins and amino acids in hydroponic fodder can accelerate muscle recovery. Quicker recovery times mean that racehorses can return to training sooner, maximizing performance in the long run.

The Unsung Hero in Racing Success

The high moisture content in hydroponic fodder can also benefit a racehorse's respiratory system, a crucial but often overlooked aspect of their health. Dust from dry feeds can irritate the horse's respiratory tract, affecting performance. Hydroponic fodder virtually eliminates this risk, supporting optimum lung function when needed most—on the racetrack.

Calming the High-Strung Athlete

Racehorses are often exposed to stress, from the noise and excitement of the racetrack to the rigors of travel. Stress can manifest in various ways. Race horses can get digestive issues and erratic behavior, impacting race performance. The B vitamins and magnesium naturally present in hydroponic fodder help to manage stress and improve the horse's focus and temperament.

Each of these elements contributes to the overall health and performance of horses in unique yet interconnected ways. By embracing hydroponic fodder as part of a holistic equine care strategy, you're pioneering a revolution in equine health and well-being.

These are a few of the health benefits for horses. A few others worth mentioning include; improved coat gloss and appearance, better behavior and temperament, stronger hooves, and earlier and more consistent heat cycles.

BEEF CATTLE

Corn [grain] produces a lower pH in the rumen, and
a lower pH promotes the growth of E. coli 0157.
Dr. William Short, Retired USDA Veterinarian

CC "Chip" Rice is a 6th-generation Texas rancher raising grass-fed Red Angus bulls in central Texas. He runs about one hundred head of cattle on close to four hundred acres. Chip claims to be the first cattleman in Texas to buy a hydroponic fodder system. Chip purchased a system to test sprouts on a small-scale before purchasing the equipment to feed his whole herd. Like me, as a younger man, Chip was first introduced to the idea of hydroponics when he viewed the EPCOT center at Disney World.

Cattle Crack

Chip states, "it took the heifers a few days to acclimate to the new feed, but now the cows are convinced that the sprouts are 'crack'." Whenever he approaches the fodder shed, the heifers sprint from wherever they are to the shed. Chip has been feeding animals for almost fifty years and has never seen animals sprint to feed before. A common theme among fodder feeders.

FodderTech reports of a third party that did a controlled feeding trial over a 12-week period with a herd of one hundred pasture-based Friesian bulls. Fifty cows were fed hay, pasture, and grain, while fifty were fed hay, pasture, and sprouts. The sprout-fed group saw a 41 percent faster daily weight gain. The feeding trial found 23 percent less dry matter consumed. Overall they saw a 27 percent lower feed cost per pound of gain. Ranchers that feed sprouts say they get a consistent flavor and better marbled in the meat.

Chip will use his existing land to increase his stocking rate by 400 percent, once he has enough hydroponic fodder capacity. Avoiding the cost of another twelve-hundred acres will in and of itself pay for the fodder system. He also believes he can change how he uses his land, such as no longer needing to put up hay.

Hydroponic Finishing and *E. coli* 0157:H7

Hydroponic fodder also offers nutritional advantages for beef cattle. As we saw in dairy cattle, fresh fodder enables the digestive system of ruminants to process food much better than grain. The natural hormones, amino acids, omega-3, beta-carotenes, and other beneficial substances increase the enzymatic activity in cows. One advantage hydroponic fodder can offer is during the finishing stage.

For example, beef cattle are traditionally finished on a corn grain-based diet. One that provides a lot of energy the last few days before slaughter, so the cow gains weight as fast as possible. Ranchers are reluctant to take their animals off corn-based feed because they see weight loss, which affects already thin margins. However, corn-based finishing diets pose unseen risks to the health of a ruminant's gut.

"You know, corn [grain] produces a lower pH in the rumen, and a lower pH promotes the growth of E. coli 0157:H7." Dr. Short begins while leaving my farm, "As a result, the lower pH injures the mucosal lining of the

15. Beef cattle eating hydroponic barley

rumen, which allows access for bacteria to enter the bloodstream. Bacteria are filtered by the kidneys and liver. During beef harvest, cattle livers are integral to making money on a beef cow."

Redefining Food Safety

Dr. William Short is a retired USDA Food Safety Inspection Service Veterinarian. The USDA employs veterinarians within the commercial slaughter industry to evaluate live and sick animals. Veterinarians ensure only safe, and wholesome food enters the food supply. So, diseased animals, organs, and carcasses are condemned.

Other than the meat we eat, beef by-products are a significant part of a rancher's revenue. Fresh fodder does not mimic the pH-lowering effects of corn grain because it is like fresh forage. As a result, organs like the liver will suffer fewer diet-related injuries. Fresh fodder gives fast, natural weight gain that results in the cow's health and well-being.

Dr. Short continues, "I have seen in the packinghouse that sometimes 80 percent of the livers would be abscessed and condemned." Abscesses were common enough that "the producers, in conjunction with the pharmaceutical companies, had a person positioned on the slaughter line watching the condemnation rate for livers." Several formulations of medicines and vaccines contain animal-derived products.

Research suggests that feeding fodder during finishing reduces the health impact on humans as well. According to Dr. Short, it also reduces E. coli 0157:H7. E. coli O157:H7 is found in the intestines of healthy cattle. Still, it causes illness to humans by eating undercooked or raw meat contaminated with the bacteria. Johns Hopkins Medicine cites that over 70,000 cases are reported to the CDC annually, and over 2,000 are admitted to medical care. Healthcare costs that can be mitigated by changing how we feed ruminants.

A Clarion Call to Action

A couple of days on a fodder-based diet would keep the rumen healthier, allow the cow to achieve weight gain goals, and reduce abscesses on organs like the liver. The animals should show us where we need to go. Overall, fodder improves animal health, increases the ability of farmers to sell by-products like liver, and reduces the impact on public health. A win-win for all. Yet, the future of hydroponic fodder in beef cattle depends on the cost and performance supplied by fodder relative to other feed supplements. At the least, fodder supplements seem like a no-brainer to any beef cattle finishing regimen.

So, the question is: Will you join this hydroponic revolution? Will you take that next step to integrate hydroponic fodder into your ranching practice? This is not just about avoiding the feds; it's about charting a new course for sustainable, humane, and economically viable beef production. The future is waiting. Are you ready?

Now, what about dairy cows?

DAIRY CATTLE

Fodder may provide some benefits on small-scale operations, farms with high land values where tillable acreage can produce high-value crops, or for producers experiencing severe, extended drought.

Soder and Heins, 2017

Hydroponic fodder isn't just a trend; it's quickly becoming a cornerstone for innovative dairy operations worldwide. For decades, traditional feed choices have been scrutinized, but hydroponic fodder brings a fresh perspective to the table—literally.

Hydroponic fodder offers a lot of benefits to your dairy animals and can overcome major constraints in producing field forage. Some of the constraints include drought, land unavailability, and your labor required for cultivation. Other factors are your pasture's growth time, the fencing you need to keep out other grazers, and natural climate calamities. Issues touched on in the previous chapter.

Farmers have virtually eliminated acidosis and laminitis, and don't need copper sulfate.

Herds see increased fertility rates and lower involuntary cull rates.

The improved herd health is seen with shinier coats and individual friendliness.[8] More healthy animals making more milk can only help the bottom line. That's because the milk is better than the milk from animals eating field forage or commercial feed.

Bradley Heins was a researcher with the University of Minnesota's Extension. From 2013-2017, he noticed the dairy industry could use some help. Studies since the early 1930s gave mixed results about feeding fodder.[9] So, Heins asked questions along the lines of, "Would feeding dairy cattle hydroponic fodder be beneficial?", and, "How?" So, what did the data show?

Follow The Data

Dairy cattle were fed a 75 percent ration of 7-day-old hydroponic corn fodder. Researchers fed corn fodder to ruminants during a 13-week feeding trial. Treatments lasted two weeks before scientists collected data. A five-day digestion trial was completed at the end of the thirteen weeks. They showed that hydroponic fodder is a good source of nutrients and has a grassroots component that helps livestock perform better.[10]

Cattle fed hydroponic fodder had an increase in the overall digestibility of crude proteins and fats. These cows had an increased yield in higher quality milk.[11] Farmers also claim they get great milk volumes from cows fed hydroponic fodder. Better digestibility and more milk lead to potentially better profits.

Beyond Quality

High-quality milk isn't just about volume; it's about nutrient density. Hydroponic fodder, rich in essential fatty acids, has improved the milk's fatty acid profile. Not only does this make for a richer, creamier product, but it also enhances the milk's nutritional value. This is paramount when consumers are becoming increasingly health-conscious and willing to pay a premium for products that offer superior nutritional benefits.

For example, FodderTech claims one of their customers started incorporating sprouts into it's ration in 2010. The cows are fed only 1 percent of their body weight as sprouts and see a 21 percent increase in revenue per CWT and 18 percent lower feed costs per cow. This results in a 17 percent Increase in milk revenue based on the feed cost ratio.[12]

The Secret Ingredient, But Just a Pinch

The properties of milk fat are largely determined by the fatty acid composition. The cow's diet significantly influences the fatty acid composition of milk. Heins confirmed that young, fresh fodder contains more fatty acids than other harvested forages when they mature. Researchers from Malta in 2019 also looked at fodder for use in dairies. They explain that the higher fat content found in milk from hydroponically fed dairy animals is likely due to a high amount of nutrients in fresh fodder.[13] This key factor could differentiate your product in a competitive market.

However, the best results were seen when cattle were supplemented with small percentages of hydroponic fodder. Therefore, feeding only hydroponic fodder to cattle is not recommended. Instead, cattle should be fed a combination of hydroponic fodder with dry fodder to promote optimal nutrient utilization.

More Data, Please

Most studies are done with traditional dairy animals like cattle, buffaloes and goats. This makes sense since these animals are the most commercially milked mammals. According to the FAO, cattle produce 81 per-

14. Holsteins eating barley

cent of world milk production, followed by buffaloes with 15 percent, goats with 2 percent and sheep with 1 percent; camels provide 0.5 percent.[14] However, these concepts extend to bison, yaks, horses, reindeers, donkeys and others. Even if humans may not drink the milk, the offspring will surely show how well hydroponic fodder is received.

Milk composition is economically important to producers and is nutritionally important to consumers. Milk is made up of fats, proteins, and water. Most of the milk fat is made up of fatty acids. Fatty acid profiles are influenced by the quality of plants used as fodder. How crops affect milk composition is of concern to feed management practices.

The Hydroponic Difference

Farmers know that there are differences in milk composition based on what the animal eats and that certain feeds make for better milk. And a savvy milk drinker can tell you what an animal ate. I prefer the cows from my local dairy, which is not the same as store bought milk.

> *"I have seen an increase in production, butterfat and protein. Now that we are feeding fodder, I have also seen the color of the milk go back to the way it was when we had pasture for the cows to graze on. Milk from pastured cows has that nice, rich, off-white color from the high levels of Vitamin A, which had been lost when they couldn't graze, and now it's back."* - Albert Pereira, Pereira Pastures Dairy

Dairy managers likely have to make many choices when deciding how to feed their herd. Hydroponic fodder production is a normal supplemental practice among the different dietary regimens. And offsetting feed costs is becoming more popular as other farmers overcome drought and resource challenges.

Hydroponic fodder is more than feed; it's a strategy for sustainable, economically viable, and high-quality dairy production. As droughts, land scarcity, and economic challenges continue to plague the industry, hydroponic fodder stands as a beacon of innovation, guiding the way toward a more sustainable and prosperous future.

SHEEP

Old-time ranchers tell me, "Before cattle ruled Colorado, sheep were king." It makes sense since sheep come in at number three—and have left a lovely hoof print on agriculture.

Their global footprint is vast. By the numbers, there are about 1.2 billion sheep worldwide, with reports of a few million more than ducks. Sheep are ruminants raised for meat, wool, skin, and milk.

Sheep wool is the most widespread and vital animal fiber in the world. Sheep are found on the expansive grasslands of northeastern China and the arid landscapes of North Africa and Central Asia. Sheep farming is also common in New Zealand, Australia, and the British Isles, areas ideal for raising these woolly animals.

Hydroponic barley, clover, and duckweed are the most studied among sheep farmers, especially in New Zealand. There, hydroponic fodder isn't just another feed option; it's a tool for enhancing the herd's overall appearance.

Hydroponic fodder maintains and often improves wool quality, enabling consistent production year-round. For instance, New Zealand farmers see improved coat appearance in their herds. Merino ewes and rams that eat clover sprouts or duckweed have better wool than sheep on forage alone.

Furthermore, lactating ewes and male calves that incorporate hydroponic barley and wheat as 25 percent of their diet exhibit healthy weight gains, positioning them as ideal candidates for wool and meat production.

GOATS

About 1 billion goats are raised around the world, primarily in China, India, Southeast Asia, and Africa. They are common in pastoralist or nomadic communities due to their ruggedness. There are now more than 300 distinct species, which shows they are bred for many different purposes. Like cattle, goats are also ruminants raised for meat, milk, fiber, leather, and labor. Milk from goats is often turned into goat cheese. Some charities provide goats to impoverished people because goats have many uses and are easier manage than cattle.

16. Goats eating barley at Hanscome Dairy

The primary reason I chose [a] fodder system was so that I could control the availability of feed for my goat herd. With our ongoing drought here in Colorado, it is difficult to find hay at all, and finding it at a reasonable price is becoming impossible. Once I've secured my barley seed inventory, I know what my fodder costs are going to be.

.Julie Hanscome, Hanscome Dairy - Kersey, CO.

Hydroponic barley, corn, and duckweed have also been fed to goats. Like the Chip's cattle, does and bucks charge the trough when the fodder arrives. Studies from Malaysia show that hydroponic corn fodder is good for goats compared to Napier grass. In a preference test, West African dwarf goats were more eager to consume fresh and dried duckweed versus Guinea grass. Dairy farmers find their goats prefer hydroponic barley, especially when hay quality sufferers from drought conditions.

Whether you're raising sheep for their prized wool or goats for their multi-purpose utility, hydroponic fodder offers a scalable, sustainable, and an economic solution for modern-day challenges in animal husbandry. In doing so, it revitalizes an age-old practice, infusing it with new life and possibilities.

PIGS

The pig is the sixth most common livestock animal. From 1960 to 2010, the number of pigs on the planet grew by 250 percent, while the size of individual pigs nearly doubled. The increased demand for animal protein contributes to the rise in pig production.[15] Pigs are also prolific eaters, have high feed conversion, and mature early. As a result, they need smaller spaces and are easier to manage.

Yet, like others, pig production is facing tremendous setbacks. The major factors responsible are the scarcity of water or land, small landholding size, and labor. Farmers struggle due to the unavailability of feed, which accounts for most of the total cost of production. Farmers using hydroponic fodder report lower feed costs.

Pigs fed hydroponic fodder as part of their diet had faster weight gain, reduced back fat and increased marbling. An organic, pastured pig farmer used modest amounts of barley sprouts over two years. They reported over 50 percent lower feed costs. They saw a 10 percent greater daily weight gain with 50 percent reduced back fat. Their hogs had 'dark/marbled meat – not light pink.'[16]

Researchers in Africa assert that pigs are the cheapest means of producing animal protein for future populations. A study in Nigeria did a feeding trial with 36 pigs. The experiment lasted for six weeks. Including hydroponic corn fodder in pig nutrition improved the performance and nutrient digestibility. In addition, there was improved health and fewer vet visits.[17]

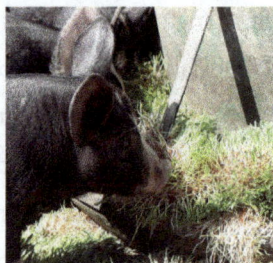

17. Pigs eating barley microfodder

The first thing that we noticed is that feeding sprouts helped heal severe foot problems. Next, the hogs at our farm weighted 20 percent more than their siblings from another farm who did not use sprouts. This was the BEST pork that I have ever eaten!

A California Hobby farmer

It is possible to feed pigs on a diet of hydroponic fodder and a conventional diet. Success shows that 50 percent hydroponic fodder is ideal. Pigs fed mixtures of concentrate and hydroponic fodder had better digestibility than feeding sprouts or traditional feed alone. [18]

By embracing hydroponic fodder, pig farmers stand to gain on multiple fronts: reduced feed costs, improved meat quality, and enhanced animal health. It's not just an alternative; it's a forward-thinking solution that could redefine the pig farming industry.

Seed Production

Something that must be of interest is the supply of seeds that are used for hydroponic fodder. Seed used in hydroponic fodder production primarily comes from soil-based farms. This division of labor allows each type of farm to specialize in what they do best—soil-based farms focus on seed production.

In contrast, hydroponic farms concentrate on generating fodder. This is familiar to farmers that grow their own field forage who may even grow their own seed. However, many buy the seed from a supplier so they can optimize space. This framework also brings about a level of dependency. What if you could grow your own seed using the same system to grow fodder, like in the old days?

The Nuances of Geographic Seed Production

Most, if not all the seeds originate in a soil farm. A farm that inevitably faces some of the challenges we already mentioned. To overcome some of those challenges, seed production is done in climates that are conducive to growth. For example, many tropical regions are dedicated to seed production for various forage crops. Corn in Hawai'i can be harvested three to four times per year and the seed is shipped back to the continental United States. Dr. Ako told me he trained students to potentially work for one the major seed companies that grow out corn seed in Hawai'i. But what if this already efficient system could be turbocharged through hydroponics?

A Hydroponic Frontier in Seed Production

Using hydroponic systems for seed production is not just innovative; it's a revolutionary idea. Hydroponics could cut many variables that make seed production challenging. No longer would farmers have to worry about unpredictable weather patterns, soil quality, or pest invasions. The hydroponic environment offers a level of control that is virtually unmatched, ensuring a consistent and reliable seed output. This innovation could lead to faster growth cycles, higher yields, and new varieties of seeds optimized for hydroponic growth.

Borrowing from Indoor Vertical Farming

For regions less endowed with abundant sunlight, the model of indoor vertical farming offers a blueprint for success. These climate-controlled facilities could run year-round, producing seeds in a staggered fashion to ensure a constant supply. Whether these seeds are then sold to other farms or used in-house for fodder production, the result is a win-win. Farmers could get more control over the supply chain.

A Vision for the Future

Hydroponic seed production extends far beyond just breaking away from traditional seed farms. It offers a vision of a more sustainable, efficient, and localized form of agriculture. Farms could become self-sufficient, producing not just fodder but also the seeds that fuel the entire operation. This model could significantly decrease the carbon emissions seen with the transport of seeds, making a huge leap to be better stewards.

In summary, the prospect of hydroponic seed production could be a paradigm-shifting idea, one that could redefine agricultural practices for years to come. It's a vision that invites further exploration, research, and, implementation. How could you be part of this groundbreaking journey?

Feeding Regimens, Small Mammals and Exotics

Hydroponic fodder studies like Heins' are limited to using one type of crop for feed. As a result, feeding solely on hydroponic fodder does not provide an animal's complete nutrition. This is in contrast to nature or standard feeding regimens. Animals frequently consume multiple types of forage in the field, and farmers vary their livestock's diets.

A combination of hydroponic fodder is likely needed to meet the nutritional demands of different species during different stages of growth. The goal is to study all the animals and mix crops in different portions, much like any animal nutritionist would do when making conventional feed. It's a bit shocking that people dismiss hydroponic fodder's viability simply based on scientific studies that control variables not likely to be seen in the field.

Exotics, Zoos and Hunters, oh my!

The most surprising industry that has adopted hydroponic fodder is the exotic. An exotic animal is loosely defined as any animal that is not a companion or used for livestock. Hydroponic fodder offers advantages for Alpaca, Elk, Deer, Buffalo, Rabbits, Tilapia, and many zoo animals like Giraffes. I was shocked to see so many giraffe show up in my search results when I typed in the words 'fodder' and 'ruminant'. It seems zookeepers know the value of hydroponic fodder.

Alpaca & Llamas farmers found faster weight gain, better coat appearance, and higher fertility. Fodder fed to rabbits showed faster weight gain and higher fertility as well.

18. Alpaca eating barley

Rabbits are popular among small acreage holders and youth agriculture programs, like 4-H. They are also raised in arid areas like the American West and Australia. An Australian rabbit farmer reports bunnies ween from mothers faster and has fewer scours when weening. The farmer exclaimed, "The Does have never been so fertile. Litter averages are up to 10 – 12 with some litters as high as 15 rabbits!"[19]

But it's not only the animals that are corralled and fed under control. A private hunting club reported feeding barley grass fodder to their deer. The deer enjoyed eating the fodder over grazing, and there were no signs of diarrhea or unused grain in their feces. Which means it was highly digestible. Hunters report a consistent taste, and some prefer to hunt on private land where deer are feed fodder rather than stalk wild deer.[20]

Hydroponic fodder is on the cusp of becoming a game-changer in the world of animal nutrition, transcending boundaries and redefining feeding regimens across a multitude of species. It's high tide we embrace this innovative approach for a more sustainable and health-conscious future.

Miyagi In Mānoa?

Feeding fish is not as simple as I assumed. My training began three months after I started school. I finally got the call from Dr. Ako. He said he had an opening in his lab and needed some help. So here was my shot at participating in some cutting-edge research, finally!

For many weeks I reported to the lab to collect a dissolved oxygen meter, plastic bottles, and a measured amount of feed. A labmate showed me how to swirl the meter in the water and scoop up the greenish water in a jar. Dr. Ako wanted to see oxygen and nutrient levels in the water, so he could be a good fish manager.

We then fed the fish, and watched them for 10 minutes. Finally, we looked at the leftovers and then recorded them in a lab notebook. The leftovers allowed us to alter the next days feeding. We were feeding fish only as much as they could eat. The wasted food would foul up the system if left unchecked. I learned how important it is to feed the proper amount. Today, I see how that training has also improved my farm's bottom line.

The first experiences working in the labs gave me the chance to learn what it means to become someone who participates in sustainable animal husbandry. Day by day, I discovered the fishes needs and learned how to raise them better. I became Dr. Ako's lead person even though I was an undergraduate. But that work came with a high cost.

Computer ON, Computer OFF

Harry soon tasked me to study the nitrification cycle and crank out weekly experiment reports. However, I first had to learn how to draft a research proposal.

As usual, I sat in his office to get daily instructions. I was told to write a proposal for a dissolved oxygen experiment. So I sauntered down to the lab, turned on the computer, opened up a blank Word document, formatted the page, and finally began to type.

After about 5 minutes, Harry came to the lab, saw me at the computer, and said, "WHAT are you doing!?" I looked at him like he was crazy and explained myself. He told me I was doing it all wrong and wasting my time. Youthful pride swelled up, and I thought, "...what the heck, I know how to use a computer, I'm fine." Instead asked, "What do you mean?"

He summoned me to his office, sat me down, pulled out a yellow legal pad, and gave me a pen. He took out another legal pad and told me to handwrite the proposal because I couldn't think on a computer. He outlined his expectations and told me I had 10 minutes.

I spent the first few minutes lamenting and hand-wrote for the rest before he summoned me again. I had little written down, but he didn't care. Another march to his office. He critiqued it and gave me 10 more minutes. I did my best and reported back.

Harry told me it was fluffy, and he wanted simple outline-type statements. I had to start over. 10 more minutes to complete the draft. I returned to his office with my latest rendition, and he told me it was much better.

Harry then explained that people rely too much on technology and waste time on computers to make things look pretty. As a result, we don't think about content and don't slow down to think. He didn't care how it looked; he wanted the information as soon as possible. Harry taught students that information now is more valuable because we can adjust in real time.

19. Sean Short and Dr. Harry Ako tend to an aquaponics system.

I try to adhere to his "down-and-dirty" philosophy every time I undertake a task. I had not taken that lesson to heart until forced to handwrite every first draft. This book was first drafted by hand. Don't tell Harry I used a tablet and a stylus, though.

Much of my education happened as an Extension Agent for Pacific farmers. I helped them put hydroponic and fish systems in place so they could grow food using less water and land. The aim was always to optimize and save farmers money. I got to do cutting-edge research and helped propel the agriculture industry. The work won an undergraduate research award in 2013 for reducing energy costs by 50 percent in an aquaponics system. I went to Hawai'i to be a fish farmer but left with a research award and a Bachelor's in Molecular Biology focused on Plant Biotechnology.

Follow The Data

What if you could make yourself a better farmer by simply tracking more of your operation? Farms that fail to keep accurate records ultimately fail. Data helps to correlate what's going on and make adjustments. Each time you collect a piece of data, you refine your process. You may not grow fodder but this one nugget could help any new farmer.

Thanks to technology, you can choose between old school and new school. Paper and pen are my first choice. Smartphones can convert to text, so I often jot something on a sticky note and capture it later. I also convert my handwritten tablet notes to text., my my second choice after pen and paper. Free software like Google Docs and Sheets allows you to easily transfer, collect and access information.

Money is an excellent example of data since most things are tied to money, and accounting software is the standard. Many farmers use software or apps like Quickbooks. That old saying "Follow the Money" can give a picture of what's happening. The flow of the money can show you what to track.

Days change with different weather, animal growth, and market changes. Thankfully we have things like the Farmers Almanac to guide us with historical weather data. The trends help us predict the future. There is also established data for various aspects of farm production. Predictions for crop and livestock yields are based on historical trends and data. And thanks to someone capturing those data, you are a better farmer.

The Main Fodder of the Book

There is a lot of evidence to support hydroponic fodder. Fodder research comes from arid regions like The Middle East, Sub-Saharan Africa, and Southeast Asia. Data also from drought-stricken areas of Australia, New Zealand, and North America. Researchers in India and Jordan have done the most work on hydroponic fodder. This should be promising, especially since we learned that the Indians contributed to the rise of modern commercial hydroponics. And Jordan is a desert country that knows how to handle drought.

Farmers also rave about fodder. For example, the hydroponic fodder manufacturer, FarmTek, ran a blog for many years highlighting customer success. Post after post talk about farmers saving money and improving their livestock quality. There is even an entire gallery of farm animals.

And the Stars Are...

The most productive hydroponic fodder crops are alfalfa, barley, and cowpea. Farmers also show that clover, corn, and sunflower are high-yielding crops. However, experience has found that barley is the highest yielding and most water efficient among the crops studied. For example, HydroGreen from Chapter One used about one-half gallon of water to produce two pounds of hydroponic barley sprouts. This is compared to the approximately twenty gallons it takes to grow the same thing in the soil.

1. StoryMaps, 2022

2. StoryMaps, 2022

3. "Nutrition", 2022

4. Alinaitwe, 2018

5. King, 2013

6. King, 2013

7. King, 2013

8. "Nutrition", 2022

9. Heins, 2016b

10. Heins, 2017

11. Heins, 2017

12. FodderTech, 2022

13. Agius et al., "Cows Fed Hydroponic Fodder and Conventional Diet: Effects on Milk Quality."

14. FAO.org, 2022

15. StoryMaps, "(Farm) Animal Planet."

16. "Nutrition.", 2022
    http://foddertech.com/nutrition/swine/

17. Adebiyi, 2018

18. Adebiyi, 2018

19. "Nutrition", 2022
    http://foddertech.com/nutrition/exotics/

20. "Nutrition", 2022
    http://foddertech.com/nutrition/exotics/

Chapter Four

THE ABC'S OF HYDROPONIC FODDER™

*The greatest service which can be rendered any country
is to add a useful plant to its culture.*

Thomas Jefferson

C rops are processed in many different ways. Most common
is to grow out the crops for their seed or grain product.
Much nutrition exists in the seed. Alternatively, crops are grown
and harvested in their green stage and are typically destined for
three different paths-silage, hay, or pasture. Since these feeds are
most common, they will be compared to hydroponic fodder.

Green forage or roughage is well known to benefit livestock.
Many farmers supplement with fresh forage because they get multiple benefits. Forage comes from the farm and improves overall

animal health. Yet, forage grown on the farm can be resource intensive.

The water plants need requires some sort of irrigation system. Soil irrigation systems can be extensive and expensive. There are entire industries dedicated to this one field. Many of these companies can benefit from the adaptation to hydroponic fodder production. However, sprinkler companies will surely disagree.

Hydroponic fodder has been extensively studied in Australia, India, and the Middle East. In addition, some work has been done in the United States. Much of the success comes from the farmer's mouth. Examples from farmers and scientists show how to consistently grow live feed on the farm. Live feed that uses less machinery and less water than field forage.

Hydroponically grown fodder saves over 90 percent of water and land. Research and farmers concurs. Unlike field production systems that use run-to-waste irrigation practices, hydroponic fodder systems can reuse water multiple times. Much like how Gericke pioneered hydroponic cultivation. Water can be recirculated through the plant roots. Many field crops or weeds that farmers use to feed livestock are candidates for hydroponic fodder. But, some crops do better than others. Seed selection is an important decision to produce the highest quality fodder and manage your farm. This chapter reveals six major crops - alfalfa, barley, clover, corn, cowpea, and sunflower.

What's There To Talk About?

These six major crops are used in field cultivation and are commonly used for animal feedstuffs. Based on the published data, we will look at the availability, yield, water use efficiency, and nutritional value of hydroponic fodder versus field forage. So, what should you grow for your animals?

Well, you grow what you already know and buy for your animals. That could be the forage seed you plant and pasture. Or it could be the hayseed or grain you harvest. Seed selection will largely depend on the animals you raise. Each crop is of importance to farmers and the larger agricultural systems.

Look to your existing supply chain. Start with a seed supplier nearby to get bulk seed. You will want to focus on pasture, reclamation, or turf seed suppliers. For example, barley and corn are used as grain, forage, and hay. Different regions will have different crop seed. So, availability will depend upon seasonal crop yields, costs, and other factors.

Prices may vary, but the more you buy, the better the deal. Small farmers may be ok with a few pounds at a time, whereas larger farms will need to buy 50 pounds or by the hundredweight. Vegetable seed companies will have some of the mentioned crops, but they may be significantly more expensive. In a pinch, I have used some or scoured Amazon. Unfortunately, I pay more than I want or get low-quality seeds.

The six crops have comparable yields to each other under hydroponic culture. This is in stark contrast to the same crops when grown in the field. Crops grown conventionally do not yield the same weight of green plant matter as hydroponic fodder.

Water use efficiency (WUE) is a ratio that measures the total amount of fodder produced given a certain amount of water used. The WUE compares how well different crops use water when grown under hydroponic or field conditions. The larger the number, the better the crop can use water. Unfortunately, there is limited data on some of the crops. However, the available data can act as a guideline for other hydroponic crops.

Nutritional properties for these crops as animal feed is available in outside literature. Yet, from a hydroponic fodder perspective, there are significant gaps with some of the crops. Therefore, more work must be done to study the nutrition of these and other crops grown under hydroponic cultivation.

Research is published in the international standard metric units. I find that many lose interest if they must do math inside their head. I have done my best to simplify that for the sake of conversation. So, I have converted metrics into the Imperial measurements used by US farmers. For instance, WUE is measured in grams or kilograms per liter of water. This means translating liters into gallons and tonnes into tons or bushels.

Alfalfa (*Medicago sativa*)

Alfalfa, the modest yet mighty crop, was where our journey at Blooming Health Farms took root. It wasn't just a plant; it was the green, leafy bridge that connected us to the local farmers and ranchers. When we first introduced our sprouted alfalfa to them, the response was like the first rain after a long drought – refreshing and full of promise.

They knew alfalfa well; it was a staple for their livestock. But our sprouted version? It was something new, something vibrant. This wasn't just about nourishing animals; it was about planting the seeds of change in the agricultural community. And so, alfalfa became much more than our first crop – it became a symbol of innovation and resilience.

—*Availability*—

Alfalfa, also called Lucerne or 'Canary clover', originated from what is now Turkey and Iran. It was likely grazed by animals long before recorded history. The first historical records of alfalfa are Turkish writings from 1300 BC. Yet, some historians claim that alfalfa cultivation began around 6,000 to 7,000 BC.

It probably was also domesticated in this area. Some historians believe alfalfa cultivation happened at the same time as the domestication of the horse. Archeological sites from the foothill of the Zargos Mountains in Turkey reveal the origins of modern agriculture. Humans needed to know what to feed their steeds, so alfalfa is probably the first plant grown for forage after the arrival of horses.

The crop is utilized worldwide and in all fifty US States. It grows in both cool and warm seasons. The plant grows to about two to three feet tall with small purple, white, or yellow flowers. Alfalfa is a legume, so it can fix nitrogen when it is inoculated with nitrogen-fixing bacteria.

Alfalfa is used in many different ways. It can be planted for pasture alone or in mixes. It also produces excellent hay. Alfalfa can grow new stems and leaves after cutting. Farmers can harvest many crops of hay in a single growing season because of this abundant regrowth.

The frequency of harvest and yields depends mainly on the growing season's length, the soil's quality, the amount of sunshine, and especially the amount of water during the growing season.

While widely used for feed, alfalfa has been shown to cause some bloating in ruminants. This is avoided through grazing management practices such as rotational grazing and feed supplementation.

Most management practices plant alfalfa in the spring, allowing farmers multiple harvests until the first frost. Since alfalfa for forage is harvested before it goes to flower, there must be enough time for the crop to complete a full growth cycle. Cutting programs vary, but alfalfa can be harvested every thirty-five to forty-five days. That means farmers can get three to five harvests per year, depending on where they are and conditions for the season.

Hydroponics alfalfa fodder can be harvested in much less time. Research shows most studies grow alfalfa fodder to a maximum of ten days. Yet, farmers have been able to produce alfalfa sprouts in a little as four days. This rapid production is the main reason the yields mentioned below are so spectacular, and for good reason.

Alfalfa is one of the most popular crops to feed livestock. Many farmer's lives and pocketbooks depends upon alfalfa. This is because fresh forage and hay are very nutritious and palatable to livestock. Alfalfa is known to contain about 16 percent protein and many mineral constituents. It is also rich in vitamins A, E, D, and K.

Those in the know will know how to guesstimate their costs based on some general information. US prices in mid-2022 place seed costs of alfalfa between $3-6 per pound. Alfalfa seed will yield about five to eight times its weight when sprouted for hydroponic fodder.

As a comparison, soil cultivation will need about six or more pounds of pure live seed per acre. $14 per forty-pound bale and $210-$250 per ton.

There's a saying that we're all just one crop away from revolutionizing our business – for us, that crop was alfalfa. As it turns out, many a farmer's livelihood hinges on the success of their alfalfa yield. When we brought our sprouted alfalfa to the market, it was like watching an old friend get a standing ovation. It was a powerful realization: it's not just about growing what we know but about growing what the market needs. This pivot wasn't just good for our business – it was good for the community and good for the land.

—Yield—

Hydroponic yields for alfalfa fodder are huge. Jordanian researchers achieved eighty-seven tons of fresh matter per acre based on an 8-day growth cycle. When dried, that equates to about seven tons of dry matter per acre.[1]

Field-grown alfalfa shows much different results. Data from the USDA's Research Service cites an average of 2.6 tons of dry matter per acre, per cutting. The record yield comes in at ten tons per acre. Averages are easy for thought, so three tons per acre seems reasonable. On average, alfalfa can be harvested every thirty-five to forty-five days, equal to three to five harvests per season. This is in line with many farmers that practice the 3-cut or 4-cut methods of alfalfa harvest. That's about nine to fifteen tons using the USDA average as an estimate.

What's to note is that field alfalfa takes thirty-five days to get an average of three tons of dry matter. Yet, hydroponic alfalfa takes only eight days to get an average of seven tons of dry matter.

For simplicity's sake, let's assume field alfalfa takes forty days before it's cut, and you get the average yield for four cuttings. Over one hundred and twenty days, a farmer will get twelve tons. A farmer growing hydroponically can grow one hundred and five tons in that same time. Now, the hydroponic fodder can be grown all year. Suppose we continue to assume the same eight-day growing cycle. In that case, the farmer can get forty-five harvests for an estimated three hundred and fifteen tons.

Earlier chapters revealed the land-sparing potentials and possible applications. I know an extrapolation can only get so much excitement. But, a twenty-five times higher potential yield should cause some pause to think about doing things differently. As Einstein reminds us, "We can't solve problems by using the same kind of thinking we used when we created them."

— Water Use Efficiency —

The same researchers from Jordan[2] looked at how much water alfalfa uses when grown hydroponically. As an arid country, Jordan is an excellent place to study the water use efficiency of crops. Hydroponic alfalfa is reported to have a WUE greater than *500* for fresh matter and about *40* for dry matter. Field Alfalfa has a WUE of *12* for fresh matter and *2* for dry matter.

Alfalfa cultivars they studied used about thirty-three hundred gallons of water to produce about one ton of hydroponic green fodder. Companies like FodderTech and HydroGreen have optimized alfalfa growout with as little as four hundred gallons of water per ton.

Field alfalfa uses about 18,500 gallons of water per ton produced. Based on the USDA average of three tons, that is 55,000 gallons per cutting and 222,000 gallons for a 120-day growing season. So the same hydroponic farmer in our pretend scenario would use 148,500 gallons of water if they grew for 360 days.

Those results mean that only about 2 percent of water is needed to produce the same amount of fodder compared to that produced under field conditions. This tremendous improvement in water use efficiency shows that hydroponic systems significantly save water to produce green fodder. The high water saving under such conditions is considered very important when a water shortage is faced. Less water with significantly higher yields.

—Nutrition—

Sprouting significantly improves the nutritional value and health quality compared to fresh alfalfa forage and hay.

In the early 2000s, Spanish researchers found the highest vitamin contents in alfalfa sprouts. They showed that hydroponic alfalfa fodder is higher in Vitamins A & C than alfalfa forage and hay. For example, the Vitamin A contents are over a thousand times higher and Vitamin C is ten times higher than in forage and hay[3].

Sprouting significantly improved the thiamine (B1) content in alfalfa which was ten times higher than in the seed. In addition, the sprouts' riboflavin (B2) contents increased four times in the alfalfa compared to the original seed. These values are in accordance with those found in the established literature.

Sprouting also increased potassium, a threefold increase in iron, and a fivefold in potassium. Hydroponic alfalfa is also higher in iron, magnesium, calcium, potassium, manganese, sodium, copper, and zinc.

Because of these results, sprouted alfalfa and other seeds are considered efficient, natural, and low-cost methods for supplying vitamins and minerals instead of fortifying foods. Also, alfalfa has the highest vitamin content of all the crops studied. However, alfalfa seed is not yet as well studied as our next crop, barley.

Barley (*Hordeum vulgare*)

As a child, I remember the earthy, rich aroma of barley that would permeate our home during my dad's beer-making sessions. It was a scent that mingled with laughter and stories, a tradition that felt as old as the crop itself. When I think of barley, I am transported to those moments, and to the local farmers that use barley in feed, which reminds me of family and community.

—Availability—

Barley is an ancient crop that originates from the Fertile Crescent and Tibet. Evidence shows that some barley cultivation began in Egypt around 5000 BC and later spread to Mesopotamia and Europe. However, recent genetic studies of ancient Egyptian barley revealed genetics from a species found in the Tibetan Plateau. Mixed genetics of barley species show humans traded and cultivated an improved barley variety for at least eight thousand years.

Globally, seventy percent of barley production is used as animal fodder. The USDA barley production for 2021 was around 2.5 million tons or 106 million bushels, down 36 percent from 2020. Data from 2022 has yet to hit the books.

Barley harvested as feed and hay is a significant source of forage for livestock producers in most arid and semi-arid regions because it can be an inexpensive and easily accessible seed. In addition, forage barley has a good yield and a higher nutritive value than other small grains.

Barley is an important feed grain in many areas with less corn production, especially in northern climates. For example, northern and eastern Europe or Russia. Barley is the main feed grain in Canada, Europe, and the northern United States. Half of the United States' barley crop is used as fodder for livestock. A finishing diet of barley is one of the defining characteristics of western Canadian beef compared to corn-finished beef in the United States.

Here in Colorado, we're proud of our title as the 'Munich of the West,' a nod to our thriving microbrew scene. Each sip of locally brewed beer is a testament to the grain that has long been a staple feed in our region, offering promise for not only our cattle but also our culture. Farmers who choose local barley varieties can enhance their technical and economic viability.

—Yield—

Most barley growers know weight by the bushel, so we will speak in both terms. For those who like mental math, there are two thousand pounds to a ton and the US Grain Council states there are forty-eight pounds per bushel of barley. For reference, a bushel is a measure of capacity equal to sixty-four US dry pints.

Hydroponic barley fodder can be produced in six to ten days from seed to harvest. Its germination time takes about one to three days. Barley yields about 190-250 tons of fresh matter per acre based on a 10-day, continuous growth cycle. When dried, that equates to about twenty-five to thirty tons of dry matter per acre. In other words, 1100-1300 bushels of barley grain equivalent.

This is in stark to field barley. Field-grown barley for feed grain is shown to yield an average of 1.25 tons of grain per acre or about fifty-two bushels. Hydroponic barley yields significantly more fresh matter and dry matter per unit area. That means we use less space to grow more barley fodder.

— Water Use Efficiency —

Researchers from Jordan who studied alfalfa above also looked at how much water barley uses when grown hydroponically. Hydroponic barley is shown to have a WUE greater than *600* for fresh matter and about *100* for dry matter. Field Barley has a WUE of *14* for fresh matter and *2* for dry matter.[4]

Barley cultivars studied used about twenty-seven hundred gallons of water to produce about a ton of hydroponic green fodder. According to Western Extension Agents in Alberta and Colorado, field barley grown as a feedstock crop needs about four to five inches of water over the crop season to produce grain. This is over 100,000 gallons of water per ton produced. Field cultivation can fluctuate water use depending on environmental conditions.

Those results mean that only about two percent of water is needed under hydroponic conditions to produce the same amount of water compared to that produced under field conditions. This tremendous improvement in water use efficiency shows that hydroponic systems significantly save water to produce green fodder for arid and semi-arid areas. The high water saving under such conditions is considered very important when a water shortage is faced. Less water with significantly higher yields.

Colorado, with its beautiful expanses of arid and semi-arid lands, knows the value of every drop of water. Our agricultural productivity might just flow as freely as our rivers, but behind every barley grain is a story of efficiency, and it's barley's ability to thrive on less water that keeps farmer's taps flowing.

—Nutrition—

Hydroponic barley is significantly more nutritious than barley grain and field forage used as feed. It has higher crude fat and protein than barley grown in the field or barley seed. The crude fat content of hydroponic barley falls between 16 percent to 25 percent, and the crude fat is 3 percent to 4 percent. This is compared to the 9 percent to 12 percent and 2 percent to 3 percent in field barley, respectively. Green fodder shows higher contents of crude protein and fat than barley and alfalfa forage grown under field conditions.[5]

The fiber contents are similar but different. There was no significant difference between hydroponic fodder and field fodder when considering crude fiber contents. The acid detergent fiber (ADF) and neutral detergent fiber (NDF) are much lower than field-grown barley. Lower ADF and NDF are considered good. Low values indicate hydroponic barley is more digestible compared to field barley. Other researchers reported that the digestibility of forages decreases with an increase in fiber content.

Initial analysis indicates that barley fodder is superior in some aspects to field-grown alfalfa or barley hay used as forage for livestock. Neutral detergent fiber (NDF) and acid detergent fiber (ADF) values for hydroponic green fodder were about thirty percent and fourteen, respectively. These fiber contents are much lower than field-grown alfalfa.[6]

Studies with livestock show promising results. Chickens had a faster weight gain and better egg laying when fed a 25 percent to 50 percent diet of hydroponic barley fodder. Broilers were brought to slaughter a week earlier than those fed a conventional diet. Layers get consistent production and omega-rich eggs.[7]

Ranchers have found similar results while milking ruminants. For example, lactating ewes fed a 25 percent diet of hydroponic barley and wheat had healthy body weight gains. Also, male calves fed hydroponic barley fodder observed healthy weight gain.[8]

In addition, researchers found higher concentrations of potassium, manganese, and zinc in hydroponically produced barley fodder than in barley and alfalfa forges produced under field conditions. Yet, minerals such as potassium, calcium, and iron are lower in hydroponic barley than in field barley. [9]

With its robust nutritional profile, hydroponic barley offers a superior feed option that supports the health of livestock and the richness of our beloved farmers and ranchers.

Clover (*Trifolium spp.*)

It was the combination of clover and alfalfa sprouts that marked my early successes in hydroponic fodder. Like siblings in agriculture, they thrived together, and their synergy was a cornerstone of Blooming Health Farms. This wasn't just about growing plants; it was about growing possibilities.

—Availability—

Clover is a legume that originates from Europe and Central Asia and is grown in most temperate and subtropical regions of the world. There are about 300 species of clover[10]. Researchers have hypothesized that the clover used in agriculture comes from species found in Africa, the British Isles, the Mediterranean, and Central Asia.

Clovers are related to alfalfa, which we we learned is sometimes called Canary clover. Clovers are small plants, typically growing up to 12 inches tall in the field. The leaves are trifoliate, with heads or dense spikes of small red, purple, white, or yellow flowers; with small, few-seeded pods.[11]

Clover is foraged by wildlife, including bears, game animals, and birds. The flowers are highly attractive to bees, and clover honey is a common secondary product of clover cultivation. Clover is useful for forage and silage for several reasons: it produces an abundant crop, is palatable to and nutritious for livestock, and is appropriate for either pasturage or green composting. Because of it's usefulness, many ancient farmers considered it the Holy Trinity[12].

Clover sprouts are popular for human consumption. Native Americans foraged for clover and ate the plants raw. Clover roots were also cooked, dried or smoked. Several species of clover are extensively cultivated as fodder plants. The most cultivated clovers varieties are white clover, *Trifolium repens*, and red clover, *Trifolium pratense*.

Most research comes from the Southern Hemisphere. Clover is an important crop to New Zealand farmers. It is a common feedstock for sheep and other pastured livestock. There is a variety of white clover called 'New Zealand Clover.' It's no surprise that a plant named after a country would also be economically important. Clover contributes a little over $3 billion to the New Zealand economy[13]. However, clover is not well studied in regard to hydroponic fodder, so more work must be done.

It's worth pondering the "good luck" often symbolized by this humble plant. In many cultures, clover is a harbinger of prosperity and good fortune. It's not just in folklore but also in practice, as integrating clover into farming practices could indeed be a fortuitous move, potentially ushering in both economic prosperity and ecological harmony.

— *Water Use Efficiency* —

The preference for clover over other crops like alfalfa or barley by water-strapped farmers is more than just a practical choice; it's a nod to serendipity. Clover has been a "lucky choice" in my own hydroponic exploits, as its water efficiency mirrors the ingenuity required in today's farming.

However, data on the water use efficiency of hydroponic clover fodder is not easily accessible, yet, we know from studies done on alfalfa and cowpea that hydroponic legumes are efficient with their use of water.

Clover research from New Zealand looked at drought-stress issues related to the entire clover industry. Researchers on the South Island needed to understand the water stress on clover so farmers could use water efficiently. They grew out varieties of white clover for thirty days to figure out which one used the least water. All the clover studied showed similar results but they revealed a WUE of 3 for dry field forage[14]. It understandable why water strapped farmers would prefer to grow clover over alfalfa or barley.

Like other legumes, clover is an ideal candidate for hydroponic fodder. Speculations seem to be in line with data provided by companies that recommend legumes as fodder for livestock.

—Yield—

Clover yields for hydroponic fodder are not readily available. However, a study done by Egyptian researchers[15] showed promising results with 3-day-old clover sprouts. Researchers got a 4-time increase in yield from seed to sprout. This is similar to other results seen with legumes like clover. Forage yields for clover can aid in understanding the green matter potential of hydroponic clover fodder. One could speculate that yields rival alfalfa when clover is grown out to a six to ten day stage.

—Nutrition—

The nutrition of hydroponic clover fodder comes from Egypt.[16] Initial studies show that hydroponic clover fodder, like alfalfa, is higher in crude protein and vitamins A, E, and C than reported values for clover forage and hay. We also know from alfalfa studies that sprouting improves legumes' nutritional value and quality compared to fresh forage and hay. Sprouting improves the B vitamin contents in sprouts compared to the seed. Hydroponic clover is likely high in B vitamins, iron, magnesium, calcium, and potassium.

The Egyptian researchers in the section on clover yield only found a 25 percent loss in dry matter with the 3-day-old clover sp routs.[17] In contrast, Jordanian researchers found a 90 percent loss of dry matter content in alfalfa microfodder.[18] While the alfalfa in Jordan was grown to 8 days, this suggests sprouts have a higher DM content than microfodder. Farmers can grow more nutritive fodder in the same amount of time because sprouts have a shorter growth cycle.

Clover sprouts studied had more crude protein than dry seeds on a dry weight basis. Egyptian researchers showed that clover sprouts had almost 54 percent crude protein. Those results are much less than the 13 percent to 25 percent protein values for clover forage from New Zealand.[19] Higher protein in hydroponic fodder is promising for feedstuff.

Even more, hydroponic clover's ash and the NDF[20] are lower than field-grown clover.[21] Lower ash and neutral detergent fiber mean the 3-day-old clover sprouts are more digestible than field forage. Beneficial to your livestock.

Highlighting the superior nutritional profile of hydroponic clover fodder, it's a good point to remind readers that for us, clover wasn't just easy to grow; it was a nutrient-packed good luck charm. In the infancy of my hydroponic journey, clover's rich content of vitamins and minerals symbolized the bounty of nature harnessed through innovation. Clover, like alfalfa, became not just a crop, but our hens' first beacon of health and vitality.

Corn (*Zea mays*)

Among the cultivated non-legume fodders, corn is the most important crop grown under irrigated conditions. It contains high concentrations of protein and minerals and possesses high digestibility. However, many of our current feed customers lean towards corn-free poultry feeds, echoing a wider conversation about sustainability and feed choice. Their preferences challenge me to think beyond traditional models and consider how hydroponic methods could reinvent corn's role in our farms.

—Availability—

Corn was born in Southern Mexico. Most historians believe corn (also known as maize) was domesticated in the south-central region of Mexico. Geneticists and ethnobotanists claim the domestication of corn started 7,500 to 12,000 years ago. Making it as old as alfalfa and barley.

There are hundreds of different varieties of corn. But in general, there are three types, dent (field), sweet, and popcorn. All the varieties will grow under hydroponic cultivation. The choice of seed can depend on cost. In most areas, dent corn will be cheaper. You may find sweet corn seed in other areas at a similar price. As sweet corn and popcorn are grown for humans, they are more likely to have a higher cost. There is little to no information on growing sweet or popcorn for fodder.

Almost every imaginable form of corn is used for feedstuff. Grain, silage, and roughage are significant components in livestock

diets, especially ruminants. It is an ideal crop as it is quick growing, high yielding, palatable and nutritious. Among the cultivated
non-legume fodders, corn is the most important crop grown under
irrigated conditions. It contains high concentrations of protein
and minerals and possesses high digestibility.[22]

Corn futures trade on several international markets, which may
give an idea of how prolific corn is as an agricultural crop. The
Encyclopedia of Life and USDA cites corn usage in the United
States for the crop year 2020 to 2021. About 40 percent of corn is
used for feed, 35 percent for industry and 10 percent for human
food, the rest destined for exports. The plastic nature of corn
makes it a promising crop beyond just fodder production.

Corn is widely cultivated worldwide, and a greater weight of
corn is produced yearly than any other grain. Many corn growers
also go by the bushel and there are fifty-six pounds per bushel
of corn. In 2021, global corn producers used almost five hundred million acres to grow 1.3 billion tons or forty-six billion
bushels.[23] Data from the USDA cites US corn producers used
eighty-five million acres to grow four-hundred-twenty million
tons, or around fifteen billion bushels. American farmers grew a
third of the world's corn harvest in an area the size of Germany.

— *Yield* —

Corn fodder can be produced in six to ten days. Hydroponic
yields about twenty-five tons of fresh matter per acre based on
an 8-day, continuous growth cycle. When dried, that equates to
about fifteen tons of dry matter per acre. This is enough to feed
thirteen to fifteen cattle daily with an average body weight gain

of 660 to 880 pounds; or 140 to 150 sheep with an average body weight gain of fifty-five to seventy-seven pounds.[24] Some farmers producing hydroponic corn fodder reported a fresh yield of almost eight pounds from one pound of corn seed in eight days. As a result, corn fodder yields five to six times more on a fresh basis than field-grown corn. When comparing corn to the field, it's best to compare it to silage and corn used for grain.

Corn is harvested at around fifty-five to sixty-five days when used as green forage. Silage-grown corn takes eighty to one hundred days. Grain-grown corn grain takes one hundred days or longer. Corn is valued for its high yield and ability to make excellent silage. It can be harvested in a single operation without significant leaf loss. Cows fed corn silage produced more milk and consumed more silage dry matter in both trials than those fed sorghum silage. Corn silage is used extensively for lactating dairy cows that need high-energy feed for maximum milk production .[25] However, hydroponic corn fodder can be ensiled just like the forage from the field.

Yet, despite a growing trend among my customers for corn-free options the yields invite us to reevaluate the benefits against environmental and health considerations, and whether hydroponic methods could provide a new narrative for corn.

— *Water Use Efficiency* —

Unfortunately, data on the water use efficiency of hydroponic corn fodder is not easily accessible. Data from Nigerian researchers is inconclusive. However, grass-like crops such as barley and corn can use water more efficiently than other plants. So, given the grass nature of corn, one could speculate a similar WUE as hydroponically grown barley. For example, field corn's water use is comparable to field-grown barley. However, corn is more water-efficient than legumes like alfalfa and forbs like sunflowers. In terms of WUE, Corn is a great candidate for hydroponic fodder.

—Nutrition—

Even though many of my feed customers specifically seek corn-free options, the nutritional benefits of hydroponically grown corn might present a case for its reconsideration, especially if it could be produced with minimal environmental impact.

Since corn is free from anti-nutritional components, it is a valuable fodder crop. Like alfalfa, corn is gluten-free and has vitamins, minerals, and antioxidants that benefit your livestock.

However, there is no easily accessible data on hydroponic corn fodder's vitamin and mineral content like there is for alfalfa and barley. Research seems to have only looked at the macromolecules in corn fodder. There is a gap to fill for vitamins and minerals. One could speculate that hydroponic corn fodder will follow the same trends as alfalfa and barley.

The crude protein content of corn seed is significantly increased after seven days of sprouting, with a dry matter content of 11 percent to 14 percent. The fiber contents are similar to field-grown corn. As with barley and alfalfa, there is also less ash. Hydroponic corn fodder contains higher crude protein, similar crude fiber, and less ash than conventional corn forage.[26]

Corn forage, silage and grain are good sources of Vitamins C, E and lutein. These vitamins promote healthy immune systems, vison and skin for your animals. Besides lutein, corn contains some important B vitamins, including niacin (B3), biotin (B7), thiamin (B2), and pyridoxine (B6).[27]

The studies done by the researchers mentioned in Chapter three showed that livestock like poultry, pigs, and ruminants benefited from a ration of hydroponic corn fodder.

Corn's presence is felt in nearly every chapter of agricultural history. However, its abundance in traditional farming contrasts with my customers' choices, who often prefer their poultry's diet corn-free. This opens a dialogue about the kind of farming we want to support and the legacy we wish to continue.

Cowpea (*Vigna unguiculata*)

Cowpea is a legacy that has journeyed through time and space from the heart of Africa to the far reaches of the globe. Its resilience mirrors the evolution of civilization itself, adapting and thriving in new environments. It's not just a plant; it's a survivor, and it carries with it the rhythm of history in its very name. It is also known as a black-eyed pea. A name that brings to mind beats and melodies that resonate through time. Much like the famous group, The Black Eyed Peas, it's all about getting things started—seeding change and watching it take root.

—*Availability*—

Cowpea is a legume from the genus *Vigna* that originates in east and southern Africa. The most diverse species are in southern Africa in the Transvaal, Cape Town, and Swaziland areas. It is one of the older domesticated crops. Remains of charred cowpeas from rock shelters in Africa have been dated to the 2nd millennium BC[28]. The name suggests it was first eaten by cattle, and today the whole plant is now as forage for animals.

Likely, the cowpea was first introduced to India and other parts of Asia when humans made tools. The cowpea might have made its way into West Africa, Southeast Asia, and parts of Europe as modern civilizations spread across the globe. Cowpea was used by the Greeks and Romans and probably reached the Americas during the 16th - 19th-century slave trading era. It has recently made its way into arid and semi-arid regions of Australia and the United States. It is also known as a black eye pea.

The only commodity of cowpea for which production estimates are available is dry grain. According to the United Nations' Food and Agricultural Organization, about four-and-one-half million tons of Cowpea grain are produced yearly on about eleven million acres worldwide. About 70 percent of cowpea production occurs in the Savanna and Sahelian parts of West and Central Africa. Areas of the world that are arid and semi-arid.[29]

Nigeria is the largest producer and consumer of grain, with approximately thirteen million acres under cultivation and an annual yields just over two million tons. After Nigeria, Niger and Brazil are the next largest producers with yearly yields of around one-half million tons. Production in the United States is about 88,000 tons, with most of the production in Texas, California, and the southern states.[30]

The crop is mainly grown for its high protein seeds, although the leaves and immature seed pods can also be consumed. The seeds can be harvested after about one hundred days. Leaves for forage be picked four weeks after planting.

—Yield—

This drought tolerance makes it an ideal crop for hydroponic fodder because it uses water efficiently. For my customers who are always looking for more sustainable ways to cultivate, this speaks volumes. The cowpea, with its humble beginnings, has become a symbol of efficiency and cultural significance, offering a path forward in a world where water is more precious than ever.

Hydroponic cowpea yields about one hundred tons of fresh matter per acre based on a 10-day, continuous growth cycle. When dried, that equates to about 15 tons of dry matter per acre.[31]

Cowpea yields of 1/2 ton per acre in Africa have been reported with only seven inches of rainfall. No other legume is capable of producing significant gains under low water conditions. This drought tolerance makes it an ideal crop for hydroponic fodder because it uses water efficiently. Cowpea varieties can produce seeds in as little as two months after planting.

— Water Use Efficiency —

One of the best next to barley. Researchers from Jordan who studied alfalfa and barley above also looked at cowpea's WUE when grown hydroponically. Hydroponic cowpea rivals barley's WUE of *600* for fresh matter and about *100* for dry matter.[32] Data for field cowpea WUE is less available. However, because cowpeas thrive in poor, dry conditions like the arid areas of Africa, they will be important during droughts as feed for livestock.

According to Jordanian researchers, cowpea uses about 2700 gallons of water to produce about 1 ton of hydroponic green fodder.[33] While field cultivation can fluctuate water use depending on environmental conditions, this drought tolerance makes it an ideal crop for hydroponic fodder because it uses water efficiently.

—Nutrition—

There is little readily available nutritional information for hydroponic cowpea. The information is limited to yield, WUE, and macronutrients from researchers in India. There is some sparse stuff that can be pieced together by looking at some other cowpea data. For instance, the nutritional properties of cowpea seed and leaves make them useful for comparison to hydroponic cowpea fodder.

The cowpea may be the "poor farmer's fodder" due to the high levels of protein found in the seeds and leaves. Cowpea seeds provide a rich source of proteins, calories, minerals, and vitamins. A seed can consist of 25 percent protein and has a very low-fat content. Cowpea seed protein content ranges from 23 percent to 32 percent of seed weight.[34] Protein isolates from cowpea grains have good functional properties. Cowpea should be looked at as a substitute for protein sources.

The crude fat content of cowpea seeds ranges from 1.4 to 2.7 percent. At the same time, fiber content is about 6 percent. What's more, cowpea starch digests slower than the starch from cereals, which is better for animal health.[35]

In addition, cowpea seeds are a rich source of minerals and vitamins and among plants have one of the highest contents of folic acid.[36] Folic acid is an essential vitamin that helps prevent birth defects in unborn animals.

The information on the nutritional value of the leaves stems primarily from researchers in Sub-Saharan Africa. A 2019 review of over eighty-six papers outlines how cowpea leaves are exploited for food and feed to reduce food and nutrition insecurity. Like the green matter of corn, they are rich in micronutrients, nutraceuticals, and antioxidants. The leaves have high levels of beta-carotene, calcium, and iron.[37] Cowpea contains essential nutrients, including vitamins and minerals, that can improve the nutritional quality of livestock.

Sunflower (*Helianthus annuus*)

The sunflower, a native of North America, holds a special place in my heart. It's not just the rapid sun-tracking that captures my attention — you can actually watch these plants grow, a rare spectacle in the plant world. As someone deeply rooted in plant science, I can't help but be mesmerized.

As I dive into the relationship between these people and the sunflower, my scientific curiosity intertwines with their spiritual connection. The sunflower's daily pursuit of the sun's light echoes the ancients' worship of solar deities — a dance of life and energy that captivates me.

—Availability—

Sunflower is a forb that originates from North America. I find it interesting that the word for 'forb' is derived from the Greek word 'phorbé', which means 'fodder'. The name 'Helianthus annuus' is derived from the Greek word 'helios' for 'sun' and 'anthos' for 'flower.' At the same time, annuus means 'annual' in Latin. A flower of the Sun that returns each year. Sunflowers are a forb grown for their edible oil and seeds.

Sunflower was cultivated by Native Americans in prehistoric North America. Evidence shows that it was first domesticated in Mexico around 2600 BC and in the southeastern US about five thousand years ago. Many indigenous American peoples used the sunflower as the symbol of their solar deity, including the Aztecs, Otomi, Hopi and the Incas.[38] In the early 1500s, early Spanish conquistadors surely saw how wildlife foraged sunflowers. Like clover, the flowers attract bees, whereas the seeds and stems are appetizing to animals. Those that left the Americas carried the sunflower seeds across the Atlantic and back to Europe.

The sunflower is currently the most economically important forb and is an important agricultural commodity. Sunflower is popular as a dual-purpose crop for both oil and forage production. Seeds of this plant are a valuable food product and raw material for the confectionery and fat-and-oil processing industries.[39] The waste from the production of sunflower oil (oil cake and meal) are in demand as feed raw materials and is siloed for silage or fed as green fodder. For example, raw feed materials are used as wild bird food, and as livestock forage. Sunflower farms find that livestock eagerly eat the leftover stems, leaves, and heads of mature plants after harvesting.

Farmers on the Great Plains have grown sunflowers for ages. Many know sunflower as a good source of feed for livestock. Some of the early research comes from North Dakota, South Dakota and Kansas. Researchers from Britain in the 1980s looked at an alternative feed aside from grain. They showed that sunflowers are a dependable source of roughage in drier climates because of their relatively high drought resistance.[40]

Pakistani researchers in the 1990s studied sheep and goats. They looked at what others were doing around the world. They proposed that sunflower plant residues could be used as a new potential roughage source for ruminants. The same scientists later showed sunflower silage in ruminant feeding optimizes feed resources, decreases feeding costs, and alleviates environmental pol lution.[41] Sunflower green matter provides good quality forage for livestock and even the environment.

In 2021, the world production of sunflower seeds was almost sixty-three million tons. Russia and Ukraine lead the world with 53 percent combined of the total. The world leaders in seed production behind the above are Argentina, India, and China. But this crop is also grown on a large scale in other countries.[42] According to the National Sunflower Association, US farmers produced around one million tons on about one-and-a-half million acres.[43] With changing world affairs, sunflower cultivation could become lucrative for some.

— *Water Use Efficiency* —

Data on the water use efficiency of hydroponic sunflower fodder is not yet available. However, as a field crop, it has been well studied.

Much WUE research is done in Texas and the Dakotas. In the early 1980s, researchers on the Great Plains paved the way for the forb's dryland destiny.[44] The dryland sunflower industry has since thrived, suggesting the crop's viability under drought-like conditions.

A Texas A&M PhD student in 2005 studied drought tolerant crops in the Texas Panhandle His dissertation aimed to understand the WUE of field-grown sorghum, sunflower, and cowpea under different tillage conditions. He found a similar WUE of *1* for sunflower and cowpea.[45] While the cowpea WUE is less than the Jordanian[46] results, it seems sunflower uses water like cowpea in similar soil conditions.

Studies in Argentina showed that the WUE of sunflower improves with increasing plant density.[47] Since fodder is grown in very high density, this is encouraging. Researchers in Turkey[48] compared the performance of different sunflowers under semi-arid conditions. They found that the variety of sunflower is a significant factor in WUE. Varieties with higher drought tolerance are recommended for sunflower fodder.

A similar WUE suggests hydroponic sunflowers will use water as efficiently as cowpea. This is promising for feedstuff since sunflower is a prolific agricultural crop. More work is needed to gather WUE for hydroponic fodder in the form of sprouts and microgreens.

The sunflower's innate ability to withstand drought conditions resonates with my own journey in hydroponics — a search for sustainable solutions in an increasingly arid world. I see a reflection of the sunflower's efficient water use in our hydroponic systems, where every drop is optimized to support growth, a principle that guides my own practices.

— *Yield* —

Sunflower yields for hydroponic fodder are not readily available. However, preliminary yields from farmers may help give an idea if it's viable. Those that have grown sunflowers are encouraged to reach out and share their successes (or failures).

—Nutrition —

Sunflower seed and oil are popular products from the sunflower plant. Unfortunately, there is less information on the nutritional qualities of sunflower fodder.

Sunflower leaves, before being ripe, are used as forage suitable for cattle and sheep. Egyptian researchers in the mid-2000s found the protein content is higher than corn leaves content, at 3.6 percent versus 1.5 percent. Higher protein in the leaves is promising for hydroponic fodder. This suggests sunflower microfodder is more nutritious than corn microfodder.[49]

Cattle farmers from Down Under reported that intercropped sunflower-corn harvested and preserved as silage is an acceptable source of forage for lactating cows. They also showed that partial replacement of corn silage by sunflower silage did not affect milk, fat, and protein yield.[50] Sunflower stalks ensiled alone or mixed with corn stalks are good roughage for ruminants. It has good palatability and adequate feeding value. It could be incorporated into lactating ruminants. Like most animal nutritionists, Turkish researchers in 2006 found better quality silage by mixing sunflower at a 50 percent ratio with other green fodder.[51] This mixed feed strategy is clearly key to any hydroponic fodder feeding strategy.

The potential of sunflower fodder to outperform traditional crops is a prospect that excites me, promising a future where our food systems are as robust as they are resource-conscious.

---

1. Al-Karaki., 2012

2. Al-Karaki., 2012

3. Plaza et al., 2003

4. Al-Karaki., 2012

5. Al-Karaki et al., 2012

6. Fazaeli et al., 2012

7. Alinaitwe, 2018

8. Saidi, 2015

9. Fazaeli et al., 2012

10. "Clover", 2022

11. "Guide To Grasses"

12. "Clover", 2022

13. Jahufer, 2021

14. Barbour et al., 1996

15. Tahany, 2018

16. Tahany, 2018

17. Tahany, 2018

18. Al-Karaki, 2012

19. O'Callaghan, 2016; Caradus, 1996

20. Tahany, 2018

21. O'Callaghan, 2016

22. Chaudhary et al., 2013.

23. Erenstein, 2021

24. Adekeye et al., 2020

25. Chaudhary, 2013

26. Naik, 2012; Lim, 2020

27. Loy and Lundy, 2019

28. Timko, 2008

29. Timko, 2008

30. Timko, 2008

31. Al-Karaki, 2012

32. Al-Karaki, 2012

33. Al-Karaki, 2012

34. Naik, 2016

35. Naik, 2016

36. Singh, 2003

37. Owade, 2019

38. Putt, 2015

39. Jones, 1984

40. Harper et al., 1981

41. Rasool, 1997

42. "Crop Explorer...", 2022

43. US Sunflower, 2021

44. (Harper et al., 1981)

45. Moroke, 2011

46. Al-Karaki, 2012

47. Echarte, 2019

48. Demir, 2020

49. Moawd, 2008

50. Moawd, 2008

51. Demiirel, 2006

Chapter Five

THE WORK INVOLVED

Your thoughts become clearer as they pass through your fingertips.

Anonymous

Action is the magic word.

Sean Short

As we learned about in the previous chapters, there are the crops and livestock. Then there is the infrastructure, labor, machinery, tools, and the desire to grow. After that is putting it into practice.

Perhaps you're inspired to grow something that uses less water or makes us more money? Perhaps you can sell the extra water? Or maybe you want to grow what you already know. The transition to hydroponics may require a deeper dive into a specific topic. The chapter will help you ask the right questions to transition to hydroponic fodder production.

We will first analyze a current or prospective operation with questions. We will then define your needs and goals by jotting down these things into an easy-to-use template. Of course, a blank sheet of paper works just as well. I often begin with a Post-it note and a pen; however, I write small.

Mentioned in this book are three different commercial fodder manufacturers. This is not a comprehensive list of individuals and companies that make fodder systems. I chose these three because they were used in the research presented, and I recommend them to farmers. Looking at other systems can help you decide if you want to pursue hydroponic fodder production.

After asking the right questions and understanding the playing field, you may need more to be successful with hydroponic fodder cultivation. The labor, equipment, infrastructure, and knowledge used in a traditional farm operation can be reduced or used similarly for hydroponic operations.

Thoughts to Ruminate

When we analyze our farm, it may be helpful to start with some questions I frequently ask to prospective hydroponic fodder farmers. Questions to ponder if hydroponic fodder is suitable for your farm. You may have already begun to ask yourself these and other questions throughout the book.

FAQs

1. What concerns do you have about water? Livestock?

2. What is your source of water?

3. What type of water rights do you have?

4. How do you use your water for your operation? How much?

5. What would it mean to you if you could use that water better?

6. If you could increase your herd with the same amount of space and water allocation, would you?

7. What do you currently feed?

8. On average, how much are you paying for your feed or forage seed to plant?

9. About how much do you currently feed (per species)?

10. What type of animals or livestock? How many?

11. How much is increasing your herd or flock worth to you?

12. How much land do you have and or use?

13. If you pasture, do you rotate? How is your growth this year? Compared to past years?

14. Do you have any certifications to be aware of? E.g., Organic or GMO-free

15. Who does the primary labor for your livestock or feed?

16. Are you willing to make a commitment to longer-term feeding - so you can see the benefits of better feed?

17. How much time do you spend on your livestock?

18. If you could get more time to do what you want, would that be important?

19. How could you better use your time?

20. What could your farm do with any money, water, time, and space you save?

21. Do you have an anticipated budget for solving your challenges?

What do You Need?

An initial analysis continues with your needs and requirements. The previous sections help answer some of these things. Beginning with what we need and defining requirements allows us to take a step back and approach things from a more impartial standpoint. Many times there are differences between what we want and what we need.

For example, we need water. But that "need" may look different for each rancher or farmer. One farmer may pull from a ditch while the other uses groundwater. One will require gates and siphons, and the other a well and pump.

What is Your Farm's Desired Outcome

Goal: _____

What specifically do you want? Describe your desired outcome in a positive sensory-based way that addresses what your outcome will do for your farm.

1. How will you know when you've achieved what you want? Determine if the "evidence" you're focused on is appropriate and timely.

2. Under what circumstances, where, when, and with whom do you want this result? Reflect on the context(s) in which you want to have this outcome and consider how getting this result may affect other areas, aspects, or people in your life.

3. What stops you from having your desired outcome already? Identify and explore any feelings, thoughts, or circumstances that seem to prevent you from your outcome.

4. What resources will you need to help you create what you want? Determine what resources you already have that will help you. Resources like time, capital, equipment, knowledge, money, connections, etc. Then, consider additional resources you'll need to move forward.

5. How are you going to get there? Identify manageable steps to help achieve your result, consider multiple options to get where you want to go, and determine the first step you'll take. Action is the magic word.

Congratulations. You just planned out your future and understand how to get where you want. Your path may be different than the next. The following sections are in no particular order. Based on what you think is your first step, I urge you to flip to the section of most interest. Now we shall talk about what you may need to get started.

Labor

The labor involved with a hydroponics operation is different but similar to labor already done on the farm. Naturally, there is time that is used to feed animals. In a traditional setting, that means giving out feed. The time spent tilling or giving out roughage is reallocated to the time spent growing and giving out fodder.

The labor required to grow fodder is minimal compared to farming field forage. Fodder is grown in a more controlled environment, unlike in the field. The controlled environment means the farmer also gets a break from the harshness of the elements.

There is an opportunity cost for everything we do. We can choose to spend that time ourselves or allocate the money and other resources to use that time as we desire. But, as farmers, we also want to work. The transition to fodder is an opportunity to transmute that desire for physical labor to something more valuable to your farm.

For instance, your farmhand is growing fodder while you're on the phone with the produce manager. You can use your time to manage or sell farm products, which is worth much more than the hourly wage you would have paid yourself to do the task done by your farmhand.

Time is the most valuable resource that we have. We must spend our time wisely and efficiently when running a farm. The time we spend allows us to do something of value. Sometimes the time is better spent doing one thing over the other. Perhaps it is more valuable for you to spend time with your animals than to negotiate with a feed salesman. The time to pursue hydroponic fodder is less than that in the field. Only a few hours a day are required to grow fodder. The level of automation you use will make this even less.

Fodder Systems

As a reminder, this is not a complete list but will serve as a good starting point. I encourage you to use your own discernment based on what learn and your situational needs.

FarmTek

The Fodder-Pro 2.0 Feed System takes away that un-certainty. Once I've secured my barley seed inventory, I know what my fodder costs are going to be. I strongly recommended this system to anyone looking at fodder systems, and will continue to do so. – Julie Hanscome, Hanscome Dairy - Kersey, CO.

FarmTek was founded in 1979 to bring high-quality products to the agricultural, horticultural, building, and retail trade industries. The About page begins, "FarmTek offers farm supplies, chicken feeders, poultry equipment, hoop buildings, and lists about twenty-seven more products and services. The company clearly supports farmers.

Since 1979, they have expanded from a two-man operation that has served over 300,000 customers worldwide. The company's current headquarters are in South Windsor, Connecticut, and they manufacture in Dyersville, Iowa. Inc. Magazine has named FarmTek one of the 500 fastest-growing companies in the United States.

Aside from the cornucopia of farm goodies, FarmTek has a few options for fodder systems. In general, they have three different sizes, depending on your needs. These are the HydroCycle Pro 9" Microgreen Table, Mini and Full Size Fodder Pro 2.0, and the Fodder Pro 3.0. Each of those sizes of systems has a limited capacity to expand before another setup is needed.

HydroCycle 9" Pro Microgreen Table

FarmTek's Microgreen Table is geared toward beginning growers or professionals. The system utilizes a recirculating nutrient film technique (NFT) and allows live-plant harvest. The small system provides microfodder with high yields. For example, sunflower shoots can yield thirteen ounces per square foot every ten days. In addition, there is less system maintenance, labor, and waste than conventional growing.

FodderPro 2.0

FodderPro 2.0 Systems are ideal for those looking to feed a few animals to larger farms with various livestock. The systems include everything a farmer needs to grow microfodder with minimal water. A trial based on barley grain found that one pound of seed can produce over seven pounds of fodder.

The systems use twelve-inch wide UV-stabilized PVC channels, which sit on a heavy-duty galvanized steel frame. The systems need about nine pounds of seed per twelve-foot channel, producing sixty to sixty-five pounds of fodder each day. Standard systems have four 12-foot-long channels per tier.

The Full-Size Systems have seven tiers of growing space, and the Mini and Micro Systems have four tiers. The company offers a multi-unit adapter kit to attach multiple units. For commercial operations with hundreds or thousands of animals and larger spaces available, the company recommends the FodderPro 3.0 Commercial Feed Modules.

FodderPro 3.0

FodderPro 3.0 Modules are ideal for large dairy operations, equine boarding facilities, commercial organic farms, beef operations, large pig farms, and commercial poultry operations. Modules sit on a galvanized steel frame and measure about six feet long. The UV-stabilized PVC channels are each sixteen inches wide and two inches deep. These systems need little maintenance and use space efficiently.

FodderPro 3.0 Commercial Feed Systems come complete with everything a farmer needs to start growing commercial-quality hydroponic feed. The efficient design of these modules allows producers to grow one ton of fodder with only one man-hour of labor each day. Sizes are incremental and combine to increase production capabilities.

FodderTech

The FodderTech shed is a one beer system. I go in to sprout with a can and by the time I am done with my chores I finish my drink.

CC "Chip" Rice

In my opinion, the system provided by FodderTech is the best and safest growing method when it comes to feeding valuable horses.

Brian Rowe, Licensed Racehorse trainer for 25 years

FodderTech develops and manufactures sprouting systems. FodderTech offers sprouting system packages to fit different livestock owners' needs and budgets. FodderTech's systems use a proprietary design that prevents mold growth. Other fodder systems and setups require operators to treat their systems with chemicals like bleach. In addition, their systems are environmentally friendly because they recycle water and fertilizer.

FodderTech offers four types of sprouting systems. These are the Table-Top and Mini, Stationary, Commercial, and Modular-Containerized systems. Unfortunately, fewer technical specifications are publicly accessible compared to FarmTek and HydroGreen.

Table-Top and Mini

Table-Top and Mini Sprouting systems are low profile and ideal for farmers with few animals. Farmers could use the system to experiment on a small scale before expanding or purchasing a larger system. The systems can sprout twenty-five to over one hundred pounds each day, depending on the model size.

Stationary

The Stationary Systems are designed for small producers and farmers that want to experiment with sprouting on a small commercial scale. All Stationary Hydroponic Sprouting systems expand to a Commercial System. Depending on size, the systems can sprout over one hundred to almost three hundred pounds each day.

Commercial

The Commercial Systems address any amount of daily production a farmer needs. Depending on the scale, these systems can produce one hundred to twenty thousand pounds each day.

Modular-Containerized

Container systems units are 100 percent turnkey and self-contained. The customer only needs a level pad, water, and electrical hookups. The systems are also scalable, producing hundreds of pounds to multiple tons of sprouts per day.

HydroGreen[1]

Imagine a fully automated hydroponic grow system on your farm that sustainably and consistently produces highly nutritious and digestible livestock feed every day while reducing greenhouse gas and using 95% less water than traditional growing methods.

HydroGreen

The company was founded by in 2015 to develop the concept of a low-maintenance feed growing system. The founder was first inspired while raising cattle in a drought-prone area of Idaho. He evolved his technology while feeding his cattle on ranches in South Dakota, Utah, and Missouri. As HydroGreen expanded, the focus was to invent a new way of growing nutritious fresh forage for animals.

Chapter one illuminated HydroGreen's contribution to hydroponic research. They also have three different sizes, depending on your needs. These are the DSG 66, GLS 808, and the Vertical Pastures™.

DSG 66

The Canadian researchers used the DSG 66 and could produce about 750 pounds of dry matter pounds of fresh forage per day. The DGS 66 machine is 6 levels high and is designed to be harvested one level each day. The DSG 66 is about eight feet by fifty feet, so it occupies about 400 square feet. One to several machines can be installed within an indoor environment. This system addresses the needs of medium-sized dairy farms and ranches with approximately one hundred to five hundred heads of cattle.

GLS 808

The GLS 808 is a larger machine with eight levels designed to be entirely harvested daily. These machines are installed within an indoor controlled environment in multiples of six. A group of six machines needs about eighty by ninety feet, occupying about seven thousand and two hundred square feet. The GLS 808 supports the feeding needs of large commercial-scale dairy farms and ranches of a thousand head or more. Six GLS 808 machines can produce about eight thousand dry matter pounds of fresh forage each day.

HydroGreen Automated Vertical Pastures™

Vertical Pastures™ are the combination of several GLS 808 machines in a commercial-scale indoor climate-controlled environment. The systems are fully automated and perform all growing functions, including seeding, watering, lighting, harvesting, and re-seeding with the push of a button. Twelve GLS 808 machines can produce twenty-five million pounds of feed grown per year, which feeds around two thousand dairy cows. A system that size occupies about a third of an acre and could replace five hundred acres of farmland.

BHF DIY Sprouting

At Blooming Health Farms, we make a variety of our own sprouting devices. Over time, we have evolved from a simple mason jar to 55-gallon volumes of sprouts. We fabricated the first ones from everyday household items before our first improvement. We used cheesecloth and window screen material to drain our soaked seeds. Our jars sat upside down on an old oven rack.

A young welder was involved with some farm startups and designed our first sprout rack. We devised a simple concept so kids who couldn't weld would have a model to build. Simple metal pieces with simple welds. Something we try to teach to our at-risk youth.

Soon after, an at-risk youth participant suggested we "glue" the screens to the mason jar bands with silicon. And like that, the BHF Sprouting Kit was born. The kit is meant for someone to sprout their own alfalfa and clover for home consumption. The product allowed us to teach hands-on entrepreneurial skills alongside farm principles. Our most recent version has a 3D-printed lid. The lid is available at www.thinkingoutsidethesoil.com or through the online shop at the branded domain. [2]

Aside from the small kit, we make larger bucket versions for in-house and farm use. We use a two-gallon and five-gallon version. The two-gallon version is used in the chapter on sprouting.

Greenhouses and Indoor Growing

Many farmers take advantage of technology to extend their growing season. Hydroponic fodder culture can be regulated. You can control the temperature and day/night cycle to improve quality. This is commonly seen as a greenhouse, barn, or another type of structure that is put up in the field to either harvest later or plant earlier in the season.

Greenhouse and indoor protection are becoming more popular in today's industrial world. This type of agriculture is best known as controlled environmental agriculture (CEA) - a way to grow the plant in the best conditions we can create. Indoor agriculture allows farmers to control their environment to achieve optimal plant growth with the ultimate goal of maximizing the resources used to be most efficient.

There are many different types of greenhouse available on the market, almost too many to even list. A brief description of some of the greenhouses I have worked with may serve as a place to start.

CERES

Ceres makes a variety of greenhouses, from DIY kits to full commercial greenhouses. They are designed in Boulder, Co, and are engineered to accommodate the colder winter temperatures. Many designs take advantage of a common practice called 'ground-air heat transfer (GAHT), where heat is alternatively stored in the ground or air to regulate the internal temperature.

Ceres partners with The Aquaponics Source to give you an out-of-the-box solution. JD and Tawnya Sawyer are considered industry leaders in aquaponics and systems support. They farming mentors for me and others and were my bosses in 2016.

GREENHOUSES IN THE SNOW

Imagine a snow-covered greenhouse in the middle of the prairie growing a banana! Well, two guys in Alliance, Nebraska, have designed a geothermal greenhouse as a heat sink. They advertise one can also grow citrus and other tropical crops like banana These greenhouses are available as kits and can be purchased for about $220 per linear foot. The shortest length you can order is fifty-four feet length, coming in around $12,000. Kits include infrastructure but do not include anything to control the climate.

FARMTEK

Sells high tunnels, cold-frames, off-the-shelf greenhouse kits, greenhouse equipment, and greenhouse irrigation. Products vary from small patio-sized sheds to buildings of hundreds of square feet. These greenhouses are fairly easy to set up.

Locate a Knowledge Base

You already research and seek information to become a better businessman. One of the resources that will help you utilize hydroponics is for you to be able to find a knowledge base. You likely have relationships with the local agricultural stores, extension agents, and others in your field. And the internet is gold.

An important thing when beginning general online research is to go past page *1* of the initial search engine inquiry. Many times the keywords used are tainted by marketing efforts. The top search results are not always the most factual.

For example, websites can be optimized for search engines based on specific keywords entered by the webmaster. The keywords may be relevant to what you're looking for but that doesn't mean that the content is appropriate for your needs. The info may be sparse or something that is popularly known. Google ranks websites by popularity and keywords.

The biggest barrier to finding good information is stopping at stuff that already confirms our beliefs. We must look for at least three things that help and three that may contradict. This causes our minds to be critical.

This is important because most claims about using hydroponics were made without validation. Then, a farmer like yourself spent countless hours on YouTube to gain a specific skill set or piece of knowledge. Time and money are important factors, and no one wants to find out they wasted them. A little time on the front end will save loads in the end.

If a crop seems to be for you, find out more about growing that crop. You are likely already an expert in growing a specific crop, and it's advisable to consider growing that plant as you do already.

One way to learn more about what you grow is to research others who are growing the same thing. I have included a list of major hydroponic operations and the types of crops they grow. The contact information listed has been obtained from the author and given permission for release. Feel free to reach out and use some of these forums for more details. The research ultimately reveals the subject matter experts in their fields. The subject matter experts are a wealth of information.

A farmer growing the same crop as you could be considered a great source of advice. One that uses hydroponic methods and offers advice, suggestions, or paid consultation. The next step would be to learn about other hydroponic fodder farmers.

This means that you have been conducting your operation with the assistance of others. That shows you already know how to find information to better your business.

Knowledge In Your Niche

Incubation

Agricultural incubator programs help farmers and locals get agricultural businesses off the ground. The staff is knowledgeable about agricultural business planning, economics, and marketing. Visits can be used to dissect and construct a custom plan for starting a farm. In fact, I used one to begin the planning for what would become Blooming Health Farms.

Most incubator programs will be associated with Land-grant Universities like Colorado State and the University of Hawai'i. Contact your state university or local Extension office to find out more.

Extension

We have mentioned Extension Agents multiple times in this book. If you have not figured it out, I advocate Extension and education as the first step to any agricultural venture. Every US State University has a College of Agriculture. These are better known as land-grant universities. Land-grant universities have a division that helps farmers with production and research needs, called Extension. In addition, each county in the United States has an Extension office with subject-matter expert agents. Agents do their best to provide resources so farmers can be successful.

Land-grant universities are traditionally used to serving farmers that have farmed the way we are all used to. Traditionally in soil. Therefore, the Agents at Extension offices throughout the US counties may have a perspective that will be absent of hydroponic principles. Those Agents that have read this book will be the first to help you. Agents in other parts of the country may be more familiar with hydroponic farming practices. With little effort, an Extension agent can understand hydroponics to facilitate a discussion.

For example, Agents in Hawai'i and Florida may have more experience with hydroponic practices. Many farmers in those states are limited by the space they can use and successfully grow hydroponic crops outdoors. This means that the Extension Agent working in those areas will likely have seen hydroponic farm operations and better understand them.

A farmer is responsible for utilizing the researchers from Extension offices and helping them understand the business. Therefore, they must network as professionals with others who know how to grow food in different ways. This leads us to a knowledge base of subject matter experts who have applied hydroponic methods.

SMEs

Within an area of knowledge, some people are experts in particular subjects. They are called subject-matter experts or SMEs (Smees, like knees with an M sound). Subject matter experts are not outright experts, but the term helps to identify their knowledge in a niche.

Many of the agents in the Weld Extension office know that I am a person interested in hydroponics, so any questions or topics concerning this fantastic way of growing plants usually come across my desk. A particular conversation about how to apply hydroponics led me to start a nonprofit farm that teaches stem principles through the medium of growing plants with water.

The Extension office received an inquiry about growing plants with fish water and forwarded that information to me. I contacted the individual to fill out an initial questionnaire to better understand his needs and what he wanted to do. When I analyzed his questionnaire, I learned he had similar aspirations as me and was trying to do something quite tricky.

Through my education, I struggled with being an at-risk youth myself. My crutch was alcohol which led me to make many poor choices. However, science and growing plants with water pulled me out of my misery and helped me pursue a path to help others.

Ryan Smith is a clinical counselor and heard about prisons in Colorado using hydroponic principles to rehabilitate prisoners. Prisoners participating in this program were less likely to return to a life of crime and become productive members of society. I recognized the similarities between my life experience and this counselor's aspirations.

In my conversation with Mr. Smith, I learned he had no experience growing plants with fish. There is a learning curve for those that have limited hydroponic agricultural expertise. I offered to help the counselor help others like him and me. Mr. Smith stepped out on a ledge to initiate an unfamiliar conversation because he had an idea to use hydroponics to solve a problem.

This book is aimed at all prospective hydroponic farmers. My conversation with Mr. Smith created Blooming Health Farms. A business that allowed this farmer to help others get their desired results.

---

1. "About", 2022

2. https://www.bloominghealthfarms.com

Chapter Six

MICROFODDER

Key Points

- Similar to sprouting

- Optimized for larger operations

- Turnover crops in one to two weeks

Microfodder

Microfodder is another name for microgreens. These are baby plants with only one to three sets of true leaves. We will use microfodder to differentiate between plants grown for livestock and microgreens grown for humans. Any sprouts you may have grown are candidates for microfodder. Many other plants can be grown as microfodder. I urge you to experiment and share your results.

Process

Step 1 – Soak and Germinate

The initial process for Microfodder is like sprouts. Depending on the volume, there are different ways to germinate. Some seed does well in a container and some seed does better when sprouted and germinated in a chosen microfodder system. For instance, a small-scale fodder system like Alyssa's mentioned below will have different needs than an extensive system like HydroGreens.

Small-scale and large-scale systems use the same principles as the chapter on Sprouts. But, you must enlarge the size of the sprouting apparatus accordingly. There are a some different strategies to do this.

Small jars of seeds like the example in the previous chapter are impractical in achieving operational efficiency. Thus, it is best to begin the soaking process with a large container like the double bucket or a barrel.

Some operators put seeds directly into tubs or trays, automating the growing process. The latter effectively germinates the sprouts in the system and makes it easy for operators on larger scales. Seed density is optimized so that there is no need to use any media for growth.

I prefer to germinate the seed before it goes into the growing area. The advantage of waiting is that you can slightly increase your yield and reduce waste. Experience has taught me that it's easier to soak seed in a container and not have it germinate rather than take up a growbed with dead seed.

Step 2 -Growout

As your seed germinates, you will notice shoots form. The shoots will begin as a yellowish shoot that turns green. At the same time, the roots will form a mat that is the foundation for the shoot's stability.

During initial growout, there is no need for lighting. Then, this may vary from crop to crop. Some scientists say that plants can wait until the fourth day. Others say day three. Many producers grow microfodder without light for up to four days. Growout generally takes six to ten days for the crops mentioned in this book.

I advocate you let the plant tell you. Some crops benefit with no light in the early days. Sunflower is a good example because it puts on a bit more plant matter as it stretches for light. As the shoots grow, you will see yellow turn to light green. This is the plant making chloroplasts to capture any available light.

Lighting may depend upon your situation. It can be as simple as sunlight or as complicated as an artificial lighting setup. Thankfully, LED technology makes artificial lighting affordable. And combined with renewable energies, artificial lighting can become a sustainable aspect of any operation.

An advantage of using light is that you can expedite your plants' growth. You can give more light, so the plant grows more. Many hydroponic operations optimize their systems by choosing the best light cycle for the crop. The more light you shine, the more sugars the plant can make to grow the roots and shoots. Of course, plants require some night, so they can rest to complete some basic biological functions, much like your livestock.

Step 3 – Harvesting

Harvesting is much like other crops in the field. You will first want to inspect for quality. Again, do a visual inspection. Smell and look at your fodder. Odor will still be the best indicator of quality. Mold smells bad. Recall the moldy bread?

Your microfodder shoots will be green, and the roots will form a dense sod-like mat, much like grass grown for a lawn. Thankfully, microfodder mats are much lighter. Rather than pour your fodder like you could have with sprouts, you will remove the mats.

Some microfodder systems have shelves that slide out. Harvest from others systems may feel a lot like shopping in a narrow warehouse aisle. Depending on the size of your fodder shelves, you may require extra hands or a knife of some sort.

You can carry Or perhaps place the strips of fodder into some type of storage or distribution vehicle so you can feed them to your animals. Some farmers have noted that their animals won't eat the roots once the shoots are grazed away. This can be overcome by shredding your microfodder before feeding. Some operators will take the mats and cut them into smaller pieces.

Companies like FodderTech sell a shredder because the shredded mats are more palatable to ruminants. Alternatively, you can compost the leftover root mass. The roots will act as a great carbon source for your soil life. Of course, it would be nice if the livestock wasn't so picky at times.

Farmers are Growing Microfodder

While working on the book, I got an email from Hannah, the Director of Extension that read,

Hi Alyssa,

Thank you for the email. I have copied Sean Short on this email who is currently looking for individuals interested in fodder systems.

It seemed a local producer I knew from the Greeley Farmer's Market named Ledingham Livestock was interested in hydroponic fodder production.

I set up a phone call, and we began our conversation with a recap of the recent County Fairs. The Ledingham kids had some small animals to show at the Larimer Fair, but they were most excited for the hoopla to be over.

Show animals are an essential part of many ranchers' lives. Kids learn how to take care of critters from a young age, teaching compassion, real-world economics, and the science of human action. Animal husbandry instills values that influence farm kids for generations. It's hard to find someone from a farm that didn't raise some animal.

After our introductory chatter, I began my fodder inquiry with some gratitude. "I'm glad I got an email from Hanna, so tell me a little bit about what it is you're looking for?"

She told me about her friends that ran Top Notch Meats. "I don't know if you know but they had to go out of business because they lost their water rights."

I did. Top Notch also wrote a letter to other vendors the previous Winter Market. It broke my heart when I heard they had to quit due to drought. That situation influenced me to help other farmers. The challenges Alyssa's friend vocalized at the beginning of this book echoed in my ear.

"I just talked to her," she continued, "I mentioned we're going to start doing fodder because we think hay prices are outrageous, and we can't sustain our livestock paying that much for hay."

"How much are you guys paying for hay?" I asked.

"I got a bill for a load of kind-of-okay hay for $7600."

"Kind of okay? Did you have to throw a bunch out?"

"No. We'll feed it out but it's not super nutritious. They're not getting a lot from it."

"I'm hearing people don't know how to hay anymore."

"Yea they really don't. We got a bunch of triticale in last year and I think it was cut too late. And the barbs on the it caused of a bunch of abscesses in the sheep's mouths. And, we had quite a few cows come up with some abscesses around their face area."

I shifted the conversation away from hay and asked about how much water they've had in Carr. The inch they had the previous week barely brought a shade of green to the grass. What could convert the sun into sugar got eaten to the ground by her sheep and cattle. Without much water, her pasture might not recover. So, Alyssa found out how to grow her own fodder.

"I went ahead and ordered the trays," she began after I asked about her fodder experience. "So I've got like 80 of those sitting in my living room right now." She then rattled off her plan to drill the trays and make her own microfodder system. But not until after Fair. Then she said something I often hear from those new to hydroponics. "I don't know. We're just kind of going at it based off of YouTube videos."

YouTube is amazing. There is a lot of valuable information online. Everyone knows something we don't. Though, sometimes there are unknown-unknowns. The unknowns we don't know about can cause the biggest barriers to business success.

Gaps are hard to fill when trying to teach ourselves new things. Because Alyssa reached out to the Extension office, she could fill in the gaps that the YouTubers left out. Yet, not every Extension Office has access to a hydroponicist.

At the beginning of the book I mentioned how I talked people out of purchasing aquaponics systems until they got more experience. The same is true with any venture into growing your own hydroponic fodder. The principles seem simple, but the systems can vary significantly in complexity. Building and operating your own system may require advice and takes time. Time is your most precious resource.

Not just to do the work, but time for what it is you need to be successful. Sometimes the people on YouTube have decades of experience behind a 10-minute video. Or the engineer has procured a supply chain that makes things cost-effective.

Alyssa and I finished the call with how she could be successful. I recommended she continue with the barley she bought and gave her the instructions from this chapter. I assured her that she could contact me and follow up with any questions. I hope to hear from her and any other farmer with water and fodder challenges.

HydroGreen

[Our] equipment can grow fresh forage in just under six days with a fraction of water used to grow alfalfa or hay on cropland

Danielle Davis, HydroGreen

HydroGreen Founder Dihl Grohs grew up in a drought-prone area of South Dakota. In 2010, he developed a low-maintenance fodder system. Growing fodder with less water was his mission.

Over eleven years later, HydroGreen systems are found on dairies and cattle ranches throughout the US and Canada.

The Wyoming Stock Growers Association recently revealed a HydroGreen system in Southeast Wyoming. A local rancher who owns seventeen thousand cattle on thirty-five thousand acres across the Western US will showcase the benefits of hydroponic fodder. Once at capacity, his system will produce eighty percent more fresh feed than HydroGreen's GLS 808 systems, providing up to seventy-two thousand pounds of fodder each day.

I learned of the project when the VP of Lending from an agricultural bank toured my humble facility. He whipped out his smartphone to show me pictures and compared my setup to one he helped finance. The banker's fingers pointed out the vertical racks while claiming an equivalent pivot irrigation system would cost almost $20 million and use a lot more water.

For example, one of HydroGreen's customers in Utah claims his eighty-acre pivot uses as much water in five days as three Hydro-Green machines will use in a year.

Like the study in chapter two, HydroGreen and the farmer will collaborate on data collection and research. The company intends to use some of the proceeds to advance the research. Further research aims to quantify the livestock feed nutrition benefits, dry matter yield percentage gain, and herd performance. There is a particular interest in fertility, milk production, and the well-being of the animals.

Chapter Seven

GREENWATER

Key Points

- Great for all sizes of operations

- Turnover duckweed crop in a few days

- Can use animal wastewater as nutrient source

Fish Make Plants Grow

E arly on in my education, Dr. Ako and I visited a farmer so we could design an crop production system. This system would use fish as the nutrient source, and the farmer wanted to grow crops that were popular among his customers.

Ed Otsuji was a tall Japanese man in his seventies. His smile and collapsed hat were the first thing I noticed. Beyond him were rows of plants that looked straight out of the movie *Jurassic Park* sprawled up a terraced section of an extinct volcano. Otsuji Farm grew Japanese vegetables that you cannot get on O'ahu on the side of Koko Head Crater.

The varieties were traditionally grown in Japan. Ed learned long ago that Japanese tourists would come to the island and buy local food to cook. Japanese people who travel do this culturally. As a younger man, he would hear from his fellow countrymen that the vegetables in Hawai'i weren't quite the same. Travelers couldn't make the same cuisines when they went to Hawai'i.

Ed started our visit with a story of how he used to raise koi fish as a youth. At first, I did not get the implication of what he was saying. He went on to tell me that he grew up that way; it was something many Japanese boys were into. Koi ponds were all over Japan. In fact, he told me, he had one right here on his farm.

I was a bit confused because this was the first time I had heard about any fish on site. Since I was in school to learn about aquaponics, I assumed that all the farmers who contacted my professor would be ignorant on the topic.

Ed walked us over to a tiny hole in the ground filled with water. I saw a skinny green hose inserted into the hole. Ed then made a few whistling sounds and said, "oh, they're about this big..." as he stretched out his arms to show me that the fish was over three feet long.

I asked Ed about his fish, and he took a scoop out of a bag of fish feed. Next, he grabbed a five-gallon bucket to remove some water and poured the water on a neighboring papaya tree. He then repeated it several times without breaking a sweat. After setting the bucket down, Ed walked over and twisted a squeaky spigot.

The hose laid on the ground like a snake and jumped when Ed turned on the water. But even after several minutes, it was hard to notice any change in the level. Water splashed from the hole as the floating feed disappeared. Still, it appeared to be reasonably clean water considering the number of fish.

Until then, I did not realize the elitist attitude that can come with academia. Thankfully, Japanese-Hawaiians have a "Wax on, Wax Off" way of teaching like in *The Karate Kid*.

Duckweed (*Lemna, Spirodela,* and *Wolffia spp.*)

—*Availability*—

Duckweeds represent a small family of aquatic floating plants consisting of at least thirty-seven species distributed all over the world.[1] Duckweeds are believed to originate in Southeast Asia, though genetic studies are not clear. These plants are the fastest-growing flowering plants and may cover ponds or lakes within a few days under favorable growth conditions. Wild animals, such as ducks, or geese, feed on duckweeds growing naturally in ponds or lakes, hence the name of the plant. Duckweed has a high nutritional value, similar to plants we've learned about already. As such, duckweeds have also been used for a long time to feed domesticated animals, either by providing them access to duckweed-grown ponds or by supplementing their diet with harvested duckweed, fresh or dried.[2]

In the 1960s and 1970s, this fodder plant was well known and used in Europe to feed waterfowl and pigs. Many researchers have recognized it as a source of protein for farm animals and aquaculture. Many studies confirm the appropriateness of feeding duckweed to farm animals: poultry – ducks, laying hens, and broiler chickens; pigs; ruminants – cattle, sheep, and goats; and fish and shrimps in aquaculture.[3]

Duckweed is a good feed supplement in diets for livestock. Among the species of duckweeds, not all are effective for protein production. In 2000, researchers assessed duckweeds to determine the species that have the greatest potential in the treatment of livestock waste and in protein production. The most common belong to the *Lemna*, *Spirodela*, and *Wolffia* genera. Researchers found that the variety *Landoltia punctata* was the best in protein production.[4] An advantage of duckweed over other protein sources is that it is characterized by better availability and absorption of amino acids, including lysine, methionine, and vitamins. It is also rich in the amino acids leucine, threonine, valine, isoleucine, and phenylalanine.

Most commercial duckweed operations are in tropical areas, where the dry meal is made after removing the oils. It is often fed to animals as natural green biomass. It grows in troves, reproducing about every three days.[5] Duckweed production can provide four to five times as much protein per area as soybean cultivation. Some other benefits of duckweed are that it is not genetically modified, contains no gluten, and requires no farmland or chemical fertilizers. Using duckweed may generate cost savings in animal production by minimizing water use and saving land.

— *Yield* —

Early studies to determine yield comes from a researcher with Brookhaven National Laboratory. In 1978, The American Scientist printed and article on the novel use of duckweed as feed and wastewater remediation. Work was done with Louisiana State University, which grew duckweed in the late 1970s.

Researchers were able to get duckweed yields of about seven to ten tons of dry matter per acre based on a continuous growth cycle. After successive research, they estimated a potential of twenty tons per acre.[6] It is of note that the results are based on a continuous growth cycle in a semi-tropical region.

However, research done most recently in 2022 has developed a sustainable duckweed system. OASIS is an experimental station in Ireland that integrates fish water and duckweed cultivation. The system uses the waste from perch and trout to grow duckweed. The duckweed is used for fertilizer and feed.

Irish researchers found that they could get a yield of almost forty-five tons of duckweed per year, twice that of the 1980s researchers. It is of note that the results are based on a continuous growth cycle in a northern latitude.

— Water Use Efficiency —

Duckweeds live on the water's surface and are effective at intercepting sunlight. Cultivation can be done in high density using wastewater, so most water loss is minimized. As a result, duckweed could be the most water-efficient plant that can be grown for fodder, however data is not easily accessible .

While aquatic, it too fixes nitrogen. The novel thing about the duckweed is that the fixed nitrogen is more available to the plants. This is because the nitrogen is immediately available in solution, unlike field alfalfa, clover, and cowpea. Promising for farmers to grow fodder with even less water and resources than any of the other ABC's of Hydroponic Fodder

—Nutrition—

Duckweeds are nutritious. The chapter, What The F*dder!?, gave some examples of livestock that benefit from duckweed. Most farmers found positive results when fed as a supplement. It is very high in crude protein and crude fats, much like other legumes. In addition, duckweed contains essential amino acids, vitamins A, B, and E, and carotenoids.[7]

Contents of protein and other ingredients in the dry matter also vary widely depending on how the duckweed is grown. Likewise, the nutrients in the wastewater may vary depending on the animal source. Livestock diets vary between species and among different practices. Manure from pigs is not the same as that from poultry. Dairy cattle eat different feed than beef cattle.

The dry matter content of duckweed is similar to other hydroponic fodder. Studies show it ranges from 3 percent to 14 percent.[8] The variation is due to the source water. Lower dry matter is associated with lower nutrient loads in the water.

The crude protein content is reported to be between about 20 percent to 45 percent. At the lower spectrum, duckweed has as much protein as forage crops. On the other side of the spectrum, duckweed may have the highest protein content of all the crops presented in this book.

The crude fat of duckweed ranged from 2 percent to 9 percent.[9] The crude fat of hydroponic duckweed is likely similar to other legumes in this book. However, trends from other legumes suggest that duckweed may have higher crude fat levels.

Duckweed also had varying levels of crude fiber at 12 percent to 28 percent.[10] As with dry matter, crude fiber contents in the duckweed may vary depending on the source water.

From a nutrition perspective, duckweed effectively absorbs various components from the aquatic environment, including harmful heavy metals. According to a 2012 FAO report, the contents of such heavy metals in duckweed do not threaten human or animal health.[11] Several studies confirm that it can be used as a supplement in feed mixtures for poultry and livestock.

Process

Step 1 – Germinate aka 'Starter Culture'

The germination of duckweed is slightly different than the previous methods and crops. As an aquatic plant, it's seed is dispersed into the water. So, the way to begin with duckweed is to get yourself a starter culture. A scoop of duckweed should be enough to get you started. The more you have, the faster you could potentially ramp up your production.

Step 2 – Growout

Unlike sprouts and microfodder in this book, duckweed will need some sort of nutrient to continuously propagate. There are a few ways to accomplish this. I use fish waste in an aquaponics system. I feed my fish and the give all my plants their nutrients. Like the researchers that used duckweed, you can use animal wastewater.

But I caution you to consult someone who can create a wastewater treatment process so you can safely manage any agricultural waste. Larger operations can use current wastewater procedures to guide them as well.

Since the duckweed is going to animals, the pathogenic risks are not likely to occur. The biggest concern with animal wastewater is making sure it is under aeration. Anaerobic conditions will promote denitrification, and you will lose the nitrogen components of your water. While that is ultimately beneficial for the environment, and a strategy used by wastewater operators, it is more beneficial for your plants to use the nitrogen.

Step 3 – Harvest

Harvesting duckweed is probably the easiest of all the methods we've discussed. Generally, there are three ways: scoop, skim, or direct access. I welcome any designs or modifications.

Scooping is the most common way of using duckweed for fodder. Many farmers use a fine-mesh net to collect as much as the net can carry. Scoops will be full of water. Livestock seems to prefer moist duckweed. Allowing the net to drain for a short time will increase the palatability of the duckweed for your livestock.

Skimmers are devices that scrape the surface of the water. They are common in wastewater treatment applications. Skimming is designed to work the duckweed without the operator. For example, large-scale duckweed producers use skimmers to remove a layer of duckweed. It effectively does the same thing as scooping.

Direct access would be a system set up for your animals to access some of the duckweed without eating the whole lot. Research showed that direct access methods need some sort of fencing or barrier to prevent overharvest and protect from the elements. My first duckweed system sat above the chicken coop. It was a good idea, until the hens thought so too.

Same Thing, Different Words

Shortly after Ed's fish demonstration, I was ready to talk about how the fish water acted as nutrients to the plants. I had long prepared to talk about the nitrification cycle. As it's better known, the Nitrogen Cycle is one of the primary biogeochemical cycles. It is intricately linked with the Water Cycle.

Both cycles are very familiar to farmers. First, nitrogen-based fertilizers are applied to the field. Then, water dissolves the fertilizer, and plants uptake what they need to grow. And crops like alfalfa and clover can fix nitrogen into the soil.

I wanted to tell Ed, "Nitrosomonas and Nitrobacter bacteria take the ammonia fish waste exuded from their gills, and convert it in an aerobic metabolic process. First, the Nitrosomonas takes the ammonia and makes nitrite. Nitrite is a toxic metabolite to fish that prevents them from using oxygen. After that, the nitrite is aerobically converted by Nitrobacter into the metabolite nitrate. Nitrate is the primary nitrogen source for the plants on your farm.

However, Ed set the bucket down and said the fish water gets turned into nitrate. He asked if I knew that smell you smell at a lake or beach. I nodded, and he said that is what the bacteria smell like. You can't see them, but they are everywhere there is water. It takes the fish waste and makes good plant food. Feed fish more, you get more plant food. In fact, he harvested algae as a soil amendment because of the high nitrates and phosphates they have.

Ed's description of the nitrification cycle told me he knew what he was talking about. He just used different words than me. Ed knew how his plants grew in the field and what to do if something changed. The algae amendments made it quite clear how he grew different crops. He even understood that the fish food was the primary nitrogen source and that by altering feed consumption, he could change the nitrogen his plants got. Brilliant!

What I quickly learned is that farmers are beyond brilliant and can be my greatest teachers. The obviousness of Ed's knowledge showed me he knew what I would talk about. He was clever because he understood the principles of what he was doing. I simply stood there with a smile and nodded my head.

---

1. Appenroth et al., 2018; Sońta et al., 2019

2. Appenroth et al., 2018; Sońta et al., 2019

3. Sońta et al., 2019

4. Appenroth et al., 2018

5. Ziegler, et al., 2014

6. Culley Jr, 1981

7. Appenroth, 2018

8. Culley Jr, 1981

9. Mohedano, 2012

10. Appenroth, 2018

11. Owen, 2010

Chapter Eight

SPROUTS

Key Points

- A fundamental skill to be mastered

- Does not need nutrients

- It can be done in less time than microgreens

It took the heifers a few days to acclimate to the new feed, but now the cows are convinced that the sprouts are 'crack'.

CC "Chip" Rice

Having "Fun" At The Y.M.C.A.

There was a time in Hawai'i while getting sober with an empty wallet when sprouting first captured my imagination. In the confined space of my YMCA dorm, a Powerade bottle became my garden. Lentil seeds, nestled within, sprouted into vibrant life, defying the limitations of space and resources. This was where I learned that with a pinch of creativity and desperation, you could grow food almost anywhere. Those sprouts were a small but potent act of rebellion against the notion that you needed a lot of space to farm. That same spirit of innovation and resourcefulness fuels every aspect of Blooming Health Farms today.

Sprouting

Sprouting is like what it sounds: germinating your seeds and then utilizing them in their sprouted state for feed. The process is simple and inexpensive.

Sprouts find their way into every meal I eat. And yes, even the ones we grow for our chickens. From the tangy burst on a slice of pizza to the subtle crunch in a spoonful of yogurt, they are my go-to garnish. This isn't just a culinary adventure; it is a personal challenge to integrate the fruits of our hydroponic labor into my daily diet, a delicious reminder of the versatility and nutritional punch of sprouts.

The nice thing about sprouts is you can use the seed you already know and have learned about. Grains, legumes, and oilseed flowers are all simple to sprout. The process is straightforward. It requires the seed of your choice, a clean freshwater source, and a little knowledge.

Sprouting is a method to increase the health qualities of food naturally. Spanish researchers we met in Chapter Four showed that sprouts have a higher nutritional value than their full-grown plant counterparts. The nutrition sections of each crop allowed us to see they contain higher levels of vitamins and minerals.

Sprouts are also a more complete protein because germination forces plants to produce all the essential amino acids necessary to create life. This means that feeding hydroponic fodder to your animals will provide many of the essential amino acids and improve the quality of their meat, fiber, or dairy product. Examples from farmers helped to show the benefits to animals bestowed by sprouts.

There are many systems available from the companies mentioned earlier. I have included systems I recommend with links in the references for easy access. The previous chapter compared some major fodder manufacturers.

You are also encouraged to use this as a guide to building a setup that makes sense for you. You can buy the materials needed to make a setup or get clever with what you have on hand. A simple design is available to download for free at www.thinkingoutsidethesoil.com

Please feel free to alter it as needed for your use. And I would love to hear the clever ways farmers can sprout. But, at the end of the day, there are many right ways.

Sprouting 101

Because our seed comes from the soil, we will want to be mindful of any type of inedible debris and molds.

Since these are naturally occurring pathogens in the stomachs of our animals, this is above and beyond what you need to do for fodder production. However, we can take some basic steps to ensure we give animals the best food possible. Our biggest concern for the food will be the mitigation of mold. Bleach is a simple, cost-effective strategies implemented into your process. Some commercial systems offer features like UV sterilization to help mitigate mold.

What follows are general instructions you can use if you want to yourself.

The BHF Sprouting Kit with raw barley will act as a guide to discuss the principles. Lids are available separately for your own jars. In addition, instructions are linked so you can make your own setups.

20. BHF Sprouting Kit

Process

We start with the seed you've selected for feed and a measured amount of dry seed. The amount will vary based on the number and type of livestock you are trying to feed. At Blooming Health Farms (BHF), we use much larger amounts of seed and devices to make feed.

Step 1 – Measure

First, take your measured amount of seed and place it into the sprouting jar. We use two dry ounces in the Sprouting Kit because the barley will fill an entire pint jar when germinated.

21. Two dry ounces of barley seed

The seed will fill varying amounts based on the type of seed. Many pint jars are marked with volumes. We put a calibration mark our jars to avoid use a measuring device each time. I have found that trying to take an exact measurement each time costs time that could be spent elsewhere. You can make a similar mark on you device with the seed once you figure out how much seed you need.

22. Two dry ounces of barley seed in jar

Add water to the jar and fill to just above the seed. You want to be sure that all the seed is submerged underwater. Some seed will float temporarily. Sunflower will float for the entire soaking process.

Step – 2 Wash

We want to wash our seed before we give it to our animals. This effectively rinses away the debris and inedible parts still on the seed. There may be some debris like dirt and plant material, depending on your seed source. Some seed is cleaner than others.

Shake or swirl to agitate the seed. You will see the water get dirty and some of debris will float. After we have rinsed our seed, it's time for the initial sanitation or soak.

Step 3a – Sanitize

A sanitation treatment is optional but recommended to eliminate any potential mold. For simplicity, we will use bleach because it is inexpensive and common for food safety. I recommend bleach made by Clorox®.

23. Barley with dirty rinse water

Contrary to widespread fears, using a mild bleach solution in sprout rinsing is safe. This practice keeps our sprouts mold-free without compromising their vitality. It offers peace of mind and a harvest we can trust.

Food safety guidelines suggest mixing no more than one tablespoon of 4.5% bleach with a gallon of water. You can prepare a stock bleach solution in a separate container. So, you soak the seeds for up to 24 hours in a two percent bleach solution.

24. *Initial barley seed soak*

Step 3b - Drain and Soak

Drain off any dirty water and rinse the seeds in water two to three times their weight. This will also get rid of the bleach if you used some. Refill the jar with fresh water. Grains will need about one-hundred-fifty percent of their volume in water. Legumes will need about two-hundred percent.

Seeds will take about four to eight hours, depending on water temperature. Cool water will take longer than warm. Be careful not to use too hot of water, because this may reduce the germination rate and encourage the growth of mold. Within twenty-four hours, seeds should have soaked up all the water.

The next step of the process is to drain the water.

25. *Soaked barley seeds after 8-hours*

Step 4 -Drain

Drain all of the water that may be left over from the initial soak. There are several ways you can dispose of this rinse water. First, you could pour it down the drain. It is similar to the wash water when rinsing your grains like rice or beans in the kitchen.

Alternatively, you could collect it for composting purposes tea. However, composting is out of the scope of this book.

Step 5 – Rinse and Repeat

After we have drained our water, we want to rinse the soaked seeds until the water is clear. As clear as you would expect from a glass of water you would drink. This will typically takes two to five complete volumes of water.

Grains are generally cleaner and require less water. However, legume crops, like alfalfa, clover, and cowpea, tend to take more of a rinsing. This happens because beans release extra proteins and sugars when they germinate. So, for those that do any legume, you may see some foam after the first soak. And don't be alarmed if your sunflower seed shells dye the water black.

You will rinse your seeds at least once daily until they are germinated. This helps them stay moist so they can sprout. You are looking for tiny white tails from the seed shells. This generally takes one to three days. Harder-shelled seeds like corn and barley may require twice a day because they hold less water.

26. Rinsed barley seed on rack to drain

The tails and green shoots help tell you the seeds are ready for the next stage.

Step 6 – Germinate

Once most of your seeds have their tails, they are ready for growth. Depending on how sprouted, the seeds may need to be transferred into a larger container like a bucket. The container can be made from a variety of farm items. The idea is to slowly drain the water when the sprouted seeds are rinsed.

26. Germinated barley seed ready for transfer

27. 2-gallon bucket with holes drilled in the bottom

The double bucket system is recommended for larger-scale sprouting and microfodder. We make larger double bucket versions for in-house and farm use, like the 2-gallon version depicted. Depending on the need, we a use a two-gallon, three-gallon, five-gallon, or 55-gallon version.

The bucket with holes is placed in another bucket without holes. Seeds are soaked in both buckets, and the top bucket can be removed to rinse. The seed stays in the buckets until they are fully sprouted.

You will continue the process for three to eight days until the seeds are fully sprouted. The sprouted seed will have a white tail and the first set of leaves. After that, the grains will look like grass seedlings, and the legumes and sunflower will have two leaves.

Once your sprouts are ready, it's time to harvest!

Step 7 - Harvest

Harvesting is much like other crops in the field. You will want to inspect the sprouts for quality. I prefer to first do a visual inspection. This is done by smelling and looking at your sprouts. The odor will be the best indicator of quality. Mold smells bad. This is the same smell you would expect from moldy bread.

Sprouts will look like healthy seeds you would start in the field. They need to be plump. Not soggy. Not too dry. The shoots will turn green, and the roots will be white. You may even see root hairs among the tangle of sprouts. In the trade, we like to say the need not be 'tired.' I think it is fascinating to see. We don't get to see the roots often unless hydroponically.

You then empty the sprouts into some type of storage or distribution device so that you can feed them to your animals.

Generally, most seeds will yield four to ten times their dry weight. For example, the two ounces of barley in our example process consistently produces one pound. That is an eight-time increase in weight.

BHF Sprouting Kits

I co-founded a nonprofit with a clinical counselor that we call Blooming Health Farms. Blooming Health Farms is a 501c3 nonprofit aquaponics and egg-laying farm in Northern Colorado that helps and employs at-risk youth between the ages of 15 and 24. We get kids off the street by empowering them with job skills, STEM education and mental health therapy in order to create the future leaders of our community.

As part of our empowerment model, we incentivize our at-risk youth with commission sales on eggs and products we make. The at-risk youth do the work needed to take care of our aquaponics and chickens. We grow leafy herbs, our own feed and better than organic eggs. We also make the things needed to help farmers do it themselves. That mostly means we make simple kits for farmers so they too can grow their own poultry feed without a lot of work or fuss. We call these our Sprouting for Success Kits.

The BHF Sprouting Kit was born out of necessity. We originally needed a simple solution to feed our backyard flock and maintain the quality. The prototype was cobbled with parts from here and there, but it worked like a charm. From those humble beginnings, we developed a kit that encapsulates the essence of Blooming Health Farms.

We have a welded stand and hold a welding class to teach simple welding techniques. We also make a 3D printed lid that shows kids the technical aspects of printing and programming. The goal is to provide skills we know are valuable for employment. Procurement, Welding, and 3D Printing are just a few examples.

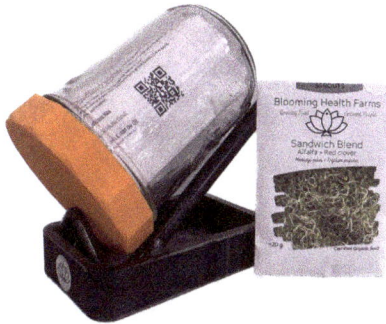

BHF Sprouting Kit

Wide-Mouth Sprouting Lid

Kewani is our hero and the at-risk youth who found his way to Blooming Health Farms mentioned in the preface to the revised edition. He became much more than just a participant in our urban farming initiative; he became a catalyst for our transformation.

His first encounter with the farm's magic was through the BHF Sprouting Kit. The kit that would soon play a significant role in not only feeding our chickens but also nourishing his potential.

The sprouting kit was straightforward, but Kewani's touch turned it into something extraordinary. The chickens thrived on this "chicken crack," and Kewani's confidence grew with every successful batch he fed out.

It was Kewani's idea, sparked by the overwhelming demand for our eggs at a local farmers market, that challenged us to think bigger. "We need more chickens," he said, planting the seed of innovation that would push us to expand just beyond the city's borders.

Together, we transformed a local couple's property into our first cooperative egg-laying operation. The sprouting kits played a pivotal role in this expansion, proving that even the simplest tools, can lead to significant advancements.

Kewani's story with the BHF Sprouting Kit is a testament to the power of opportunity and the impact of hands-on learning. His journey from a curious newcomer to a key member of our team embodies the essence of Blooming Health Farms. And a tribute to our tagline, Growing Food. Growing People.

Find the kits and lids for your own jars at
https://www.bloominghealthfarms.com/store/

Epilogue

Principles over particulars

Sean Short

One hundred years ago, most of us farmed and grew our own food. Today, less than three percent of Americans make the things that grow our great nation. But thanks to advances in technology, business, and science, we've stretched those numbers, doing much more with much less.

The drought isn't just an occasional nuisance; it's a life-altering catastrophe for farmers. This isn't about sporadic dry spells anymore. We're talking about aridification, the act of turning from a dry area into a desert. Water has become our most precious resource, and the time to use it wisely is now.

As world populations are set to boom beyond 10 billion, experts are sounding alarms that there won't be enough arable land to sustain us all. But these grim forecasts often overlook one crucial element: the proven potential of hydroponic fodder. With hydroponics, we can save over 90 percent of the land and water used in traditional agriculture. Research and farmers alike concur on this revolutionary shift.

Not only does hydroponic farming conserve resources, but it's also a gift to Mother Earth. Imagine sparing vast stretches of land, transforming them into agricultural land trusts and natural preserves. We're not just talking about fewer inputs; we're talking about a seismic shift in how much food can be produced.

Farmers and ranchers are the oldest of pioneers, bearing the weight of centuries of challenges. Isn't it high tide we pivot to a path that's not just new but has been trailblazed and proven? Hydroponic agriculture has sustained human life for thousands of years, and it's begging to be scaled.

Farmers like you have been changing the world for generations. We rely on your ingenuity to defy the doomsayers. Books have been written, talks have been given, but let's be honest—action is the magic word.

As the great R. Buckminster Fuller said,

> *We cannot fight the existing model by trying to change it, but we can introduce a new model that makes the old model obsolete.*

About Author

A young aquaponicist and Chicken Pimp revolutionizing how to think about food, education, and criminal justice.

Thirteen years of hydroponics and water treatment experience. Skilled in wet chemistry, biochemistry, and molecular biology lab techniques, as well as technical skills involved in plant science and quality management.

Received an undergraduate research award for optimizing an aquaponics system in 2013, leading to a fifty percent reduction in energy costs. Spent time in the Denver-Julesburg basin as a Wastewater Plant Operator and a Wastewater Equipment Applications Engineer across the Midwest.

Co-founder and Executive Director of Blooming Health Farms. Blooming Health Farms is a 501c3 nonprofit aquaponics and chicken egg-laying farm in Northern Colorado that empowers at-risk youth between the ages of 15 and 24. We get kids off the street by teaching job skills, STEM education, and mental health support in order to create contributing members of our community.

Currently appointed as the Co-Chair and Horticulture Advisor to Weld County Extension and pursuing a MSEng at Johns Hopkins in Systems Engineering.

Learn more at https://www.bloominghealthfarms.com
Contact Sean Short
sean@thinkingoutsidethesoil.com
OR
sean@bloominghealthfarms.com

1. "11. White Clover." Agronomy Society of New Zealand. Accessed July 1, 2022. https://www.agronomysociety.o rg.nz/11.-white-clover.html.

2. "About." HydroGreen, April 26, 2022. https://hydrogr eenglobal.com/about/.

3. Adebiyi, O. A. "Effects of Feeding Hydroponics Maize Fodder on Performance and Nutrient Digestibility of Weaned Pigs." Applied Ecology and Environmental Research 16, no. 3 (2018): 2415–22. https://doi.org/10.15 666/aeer/1603_24152422.

4. Adekeye, Adetayo Bamikole, Olufemi Sunday Onifade, Goke Tunde Amole, Ronke Yemisi Aderinboye, and Olufunmilayo Alaba Jolaoso. "Water Use Efficiency and Fodder Yield of Maize (Zea Mays) and Wheat (Triticum Aestivum) under Hydroponic Condition as Affected by Sources of Water and Days to Harvest." African Journal of Agricultural Research 16, no. 6 (2020): 909–15. http s://doi.org/10.5897/ajar2019.14503.

5. Agius, Alan, Grazia Pastorelli, and Everaldo Attard. "Cows Fed Hydroponic Fodder and Conventional Diet: Effects on Milk Quality." Archives Animal Breeding 62, no. 2 (2019): 517–25. https://doi.org/10.5194/aab-62-5 17-2019.

6. Alinaitwe, Justine, A. S. Nalule, S. Okello, S. Nalubwama, and E. Galukande. "Nutritive and economic value of hy-

droponic barley fodder in kuroiler chicken diets." IOSR J. Agric. Vet. Sci 12 (2018): 76-83.

7. Al-Karaki, Ghazi N., and M. Al-Hashimi. "Green Fodder Production and Water Use Efficiency of Some Forage Crops under Hydroponic Conditions." ISRN Agronomy 2012 (2012): 1–5. https://doi.org/10.5402/2012/9 24672.

8. "Alfalfa." Wikipedia. Wikimedia Foundation, January 1, 2022. https://en.wikipedia.org/wiki/alfalfa.

9. "Animal Products." USDA ERS - Animal Products. Accessed June 30, 2022. https://www.ers.usda.gov/topics/animal-products/.

10. Appenroth, Klaus-J., K. Sowjanya Sree, Manuela Bog, Josef Ecker, Claudine Seeliger, Volker Böhm, Stefan Lorkowski, et al. "Nutritional Value of the Duckweed Species of the Genus Wolffia (Lemnaceae) as Human Food." Frontiers in Chemistry 6 (2018). https://doi.org/10.3389/fchem.2018.00483.

11. Barbour, M., J.R. Caradus, D.R. Woodfield, and W.B. Silvester. "Water Stress and Water Use Efficiency of Ten White Clover Cultivars." NZGA: Research and Practice Series 6 (1996): 159–62. https://doi.org/10.33584/rps.6 .1995.3359.

12. "Barley." Wikipedia. Wikimedia Foundation, January 1, 2022. https://en.wikipedia.org/wiki/barley.

13. Barlow, Maude. Blue Covenant: The Global Water Crisis and the Coming Battle for the Right to Water. New York, NY: New Press, 2009.

14. "Bushel / Tonne Converter." Agriculture, Forestry and Rural Economic Development : Applications & Tools. Government of Alberta. Accessed September 4, 2022. https://www.agric.gov.ab.ca/app19/calc/crop/bu shel2tonne.jsp.

15. Callaway, Todd R., M. A. Carr, T. S. Edrington, Robin C. Anderson, and David J. Nisbet. "Diet, Escherichia coli O157: H7, and cattle: a review after 10 years." *Current issues in Molecular Biology* 11, no. 2 (2009): 67-80.

16. Caradus, J.R., D.R. Woodfield, and A.V. Stewart. "Overview and Vision for White Clover." NZGA: Research and Practice Series 6 (1996): 1–6. https://doi.or g/10.33584/rps.6.1995.3368.

17. Chakrabarti, Rina, William D. Clark, Jai Gopal Sharma, Ravi Kumar Goswami, Avanish Kumar Shrivastav, and Douglas R. Tocher. "Mass Production of Lemna Minor and Its Amino Acid and Fatty Acid Profiles." Frontiers in Chemistry 6 (2018). https://doi.org/10.3389/fchem.20 18.00479.

18. Chaudhary, D. P., S. L. Jat, R. Kumar, A. Kumar, and B. Kumar. "Fodder Quality of Maize: Its Preservation." Maize: Nutrition Dynamics and Novel Uses,

2013, 153–60. https://doi.org/10.1007/978-81-322-16
23-0_13.

19. Chavan, J. K., S. S. Kadam, and Larry R. Beuchat. "Nutritional Improvement of Cereals by Sprouting." Critical Reviews in Food Science and Nutrition 28, no. 5 (1989): 401–37. https://doi.org/10.1080/10408398909527508.

20. Chen, Zhikun, Chunjiang An, Hanxiao Fang, Yunlu Zhang, Zhigang Zhou, Yang Zhou, and Shan Zhao. "Assessment of regional greenhouse gas emission from beef cattle production: a case study of Saskatchewan in Canada." *Journal of Environmental Management* 264 (2020): 110443.

21. Ciardullo, Joshua. Hydroponic Fodder Needs Assessment and Ranch Tour. Personal, July 25, 2022.

22. ClearSpan Fabric Structure, ed. "Alpaca Lunch – a Fodder Lunch, That Is!" FarmTek Blog, July 30, 2013. https://farmtek.wordpress.com/2013/07/29/fodder_for_alpacas/.

23. ClearSpan Fabric Structure, ed. "Spotlight on Pereira Pastures Dairy – Less Pasture? More Fodder!" FarmTek Blog, September 17, 2013. https://farmtek.wordpress.com/2013/09/17/pereira_pastures_dairy/.

24. "Clover." Wikipedia. Wikimedia Foundation, January 1, 2022. https://en.wikipedia.org/wiki/clover.

25. "Cowpea." Wikipedia. Wikimedia Foundation, January 1, 2022. https://en.wikipedia.org/wiki/Cowpea.

26. "Crop Explorer for Major Crop Regions - United States Department of Agriculture." International Production Assessment Division (IPAD) - Home Page. Accessed August   2,   2022. https://ipad.fas.usda.gov/cropexplorer/cropview/comm odityView.aspx?startrow=11&cropid=2224000&sel_yea r=2021&rankby=Production.

27. Crothers, Laura. "White Clover." Sustainable Agriculture Research & Education Program, March 30, 2021. https://sarep.ucdavis.edu/covercrop/whiteclover.

28. Culley Jr, Dudley D., Eliska Rejmánková, Jan Květ, and J. B. Frye. "Production, chemical quality and use of duckweeds (Lemnaceae) in aquaculture, waste management, and animal feeds." Journal of the World Mariculture Society 12, no. 2 (1981): 27-49.

29. Davis, Alyssa. "Spotlight on Hanscome Dairy and Why They Feed Fodder." FarmTek Blog, May 28, 2013. https://farmtek.wordpress.com/2013/05/28/spo tlight-hanscome-dairy/.

30. DEMİR, İsmail. "Comparing The Performances of Sunflower Hybrids in Semi-Arid Condition." Türk Tarım ve Doğa Bilimleri Dergisi 7, no. 4 (2020): 1108-1115.

31. Demiirel, Murat, Duran Bolat, Sibel Celik, Yunus Bakici,

and Ahmet Tekeli. "Evaluation of Fermentation Qualities and Digestibilities of Silages Made from Sorghum
and Sunflower Alone and the Mixtures of Sorghum-Sun-
flower." Journal of Biological Sciences 6, no. 5 (2006):
926–30. https://doi.org/10.3923/jbs.2006.926.930.

32. Despommier, Dickson D. The Vertical Farm: Feeding the
World in the 21st Century. New York, NY: Picador, 2011.

33. Douglas, James Sholto. Hydroponics: The Bengal System. 3rd ed. Bombay: Indian Branch, Oxford University
Press, 1959.

34. "Drought." Drought.gov. Accessed August 3, 2022. htt
ps://www.drought.gov/.

35. Ebel, Roland. "Chinampas: An Urban Farming Model of
the Aztecs and a Potential Solution for Modern Mega-
lopolis." HortTechnology 30, no. 1 (2020): 13–19. htt
ps://doi.org/10.21273/horttech04310-19.

36. Egyptian Hieroglyph. Egypt and Aquaculture/Hydro-
ponics. Institute of Simplified Hydroponics. Accessed
February 2, 2021. http://www.carbon.org/school/newc
lass/egypt.htm.

37. El-Deeba, Mona. M., Mohamed N. El-Awady, Mahmoud
M. Hegazi, Fathy A. Abdel-Azeem, and Mahmoud M.
El-Bourdiny. "Engineering Factors Affecting Hydroponics Grass- Fodder Production." Misr Journal of Agricultural Engineering 26, no. 3 (2009): 1647–66. https://do

i.org/10.21608/mjae.2009.108766.

38. EPA. Environmental Protection Agency. Accessed July 1, 2022. https://www.epa.gov/watersense/how-we-use-water.

39. Erenstein, Olaf, Jordan Chamberlin, and Kai Sonder. "Estimating the Global Number and Distribution of Maize and Wheat Farms." Global Food Security 30 (2021): 100558. https://doi.org/10.1016/j.gfs.2021.100558.

40. "Escherichia Coli O157:H7." Johns Hopkins Medicine, August 14, 2019. https://www.hopkinsmedicine.org/health/conditions-and-diseases/escherichia-coli-o157-h7.

41. "Fao.org." Dairy production and products: Dairy animals. Accessed August 4, 2022. https://www.fao.org/dairy-production-products/production/dairy-animals/en/.

42. "Farmtek Blog." FarmTek Blog. Accessed July 30, 2022. https://farmtek.wordpress.com/.

43. Farmtek. Accessed August 7, 2022. https://www.farmtek.com/farm/supplies/AboutUsView.

44. Fazaeli, H., H. A. Golmohammadi, S. N. Tabatabayee, and M. Asghari-Tabrizi. "Productivity and nutritive value of barley green fodder yield in hydroponic system." World Applied Sciences Journal 16, no. 4 (2012): 531-539.

45. Fazaeli, Hassan, Somayeh Solaymani, and Yousef Rouzbahan. "Nutritive Value and Performance of Cereal Green Fodder Yield in Hydroponic System." Research on Animal Production 8, no. 15 (2017): 96–104. https://doi.org/10.29252/rap.8.15.96.

46. Feed Solutions, LLC. "Automated Vertical Pastures®." Feed Solutions, LLC, 2022. https://feedsolutionsllc.com/hydrogreen-systems.

47. Gericke, W. F. "On the physiological balance in nutrient solutions for plant cultures." American Journal of Botany 9, no. 4 (1922): 180-182.

48. Gericke, W. F. "Aquaculture, a means of crop production." American Journal of Botany 16 (1929): 862.

49. "Goat Overview | Attra | Sustainable Agriculture Project," August 2004. https://attra.ncat.org/goatoverview/.

50. Goddek, Simon, Alyssa Joyce, Benz Kotzen, and Gavin M. Burnell. Aquaponics Food Production Systems Combined Aquaculture and Hydroponic Production Technologies for the Future. Cham: Springer International Publishing, 2019.

51. Harper, F., Elizabeth Donaldson, Annie R. Henderson, and R. A. Edwards. "The potential of sunflower as a crop for ensilage and zero-grazing in northern Britain." The Journal of Agricultural Science 96, no. 1 (1981): 45-53.

52. HARRIS, SHARON L., MARTIN J. AULDIST, DAVID A. CLARK, and ERNA B. JANSEN. "Effects of White Clover Content in the Diet on Herbage Intake, Milk Production and Milk Composition of New Zealand Dairy Cows Housed Indoors." Journal of Dairy Research 65, no. 3 (1998): 389–400. https://doi.org/10.1017/s00 22029998002969.

53. Hassell, Wendell G. Guide To Grasses. Greeley, Co: Pawnee Buttes Seed, Inc., 2016.

54. Heins, B. J., J. Paulson, and H. Chester-Jones. "0662 Evaluation of Production, Rumination, Milk Fatty Acid Profile, and Profitability for Organic Dairy Cattle Fed Sprouted Barley Fodder." Journal of Animal Science 94, no. suppl_5 (2016a): 316–17. https://doi.org/10.2527/jam2016-0662.

55. Heins, Bradley. "Evaluation of fodder systems for organic dairy cattle to improve livestock efficiency." Ceres Trust. (2016b).

56. Hillman, William S., and Dudley D. Culley. "The Uses of Duckweed: The Rapid Growth, Nutritional Value, and High Biomass Productivity of These Floating Plants Suggest Their Use in Water Treatment, as Feed Crops, and in Energy-Efficient Farming." American Scientist 66, no. 4 (1978): 442–51. http://www.jstor.org/stable/27848752.

57. Hobbs, Greg, and Michael E. Welsh. Confluence: The

Story of Greeley Water. Greeley, CO: City of Greeley, 2020.

58. "Home: Food and Agriculture Organization of the United Nations." FAO Home. Accessed June 21, 2022. http s://www.fao.org/.

59. "How Our Food System Is Destroying the Nation's Most Important Fishery." Grist, May 18, 2012. https://grist.o rg/article/2010-02-08-who-owns-the-dead-zone/.

60. "Hydrogreen Offers Solutions." Wyoming Livestock Roundup, July 1, 2022. https://www.wylr.net/2022/07 /01/hydrogreen-offers-solutions/.

61. Hydroponic Fodder Systems. Accessed June 30, 2022. http://foddertech.com/.

62. "Hydroponics History Part 2: The Birth of Hydroponics." Hydroponics, September 25, 2016. http://hydrop onicgardening.com/history-of-hydroponics/.

63. Jahufer, M. Z., J. L. Ford, G. R. Cousins, and D. R. Woodfield. "Relative Performance of White Clover (Trifolium Repens) Cultivars and Experimental Synthetics under Rotational Grazing by Beef Cattle, Dairy Cattle and Sheep." Crop and Pasture Science 72, no. 11 (2021): 926. https://doi.org/10.1071/cp21084.

64. Jones, Ordie R. "Yield, Water-Use Efficiency, and Oil Concentration and Quality of Dryland Sunflower Grown

in the Southern High Plains 1." Agronomy Journal 76, no. 2 (1984): 229–35. https://doi.org/10.2134/agronj1 984.00021962007600020014x.

65. Kannan, M, G Elavarasan, A Balamurugan, B Dhanusiya, and D Freedon. "Hydroponic Farming – a State of Art for the Future Agriculture." Materials Today: Proceedings, 2022. https://doi.org/10.1016/j.matpr.2022.08.416.

66. Kim, Dong-Kwan, Young-Min Kim, Sang-Uk Chon, Kyung-Dong Lee, and Yo-Sup Rim. "Growth Characteristics and Nutrient Content of Cowpea Sprouts Based on Light Conditions." The Korean Journal of Crop Science 60, no. 4 (2015): 475–83. https://doi.org/10.7740/kjcs. 2015.60.4.475.

67. King, Kerry. "It's a Horse, of Course – so Feed It Fodder!" FarmTek Blog, May 29, 2013. https://farmtek.wordpress.com/2013/05/29/its -a-horse-of-course-so-feed-it-fodder.

68. Kossiakoff, Alexander. *Systems Engineering: Principles and Practice.* Hoboken: J. Wiley & Sons, 2020.

69. Ledingham, Alyssa. Hydroponic Fodder Needs Assessment and Assistance. Phone Interview, August 3, 2022.

70. Lennard, Wilson, and Simon Goddek. "Aquaponics: the basics." *Aquaponics Food Production Systems* (2019): 113-143.

71. "Lemna Minor." Wikipedia. Wikimedia Foundation, January 1, 2022. https://en.wikipedia.org/wiki/Lemna_minor.

72. Lim, Whay Chuin, Mohd Noor Hisham Mohd Nadzir, Mark Wen Han Hiew, Md Mamat, and Shamarina Shohaimi. "Feed Intake, Growth Performance and Digestibility of Nutrients of Goats Fed with Outdoor-Grown Hydroponic Maize Sprouts." Pertanika Journal of Tropical Agricultural Science 45, no. 1 (2022).

73. Loy, D.D., and E.L. Lundy. "Nutritional Properties and Feeding Value of Corn and Its Coproducts." Corn, 2019, 633–59. https://doi.org/10.1016/b978-0-12-811971-6.00023-1.

74. "Man Jumps From Beaver Run Roof." Summit Daily. January 19, 2004. https://www.summitdaily.com/news/man-jumps-from-beaver-run-roof/.

75. Merriam-Webster's Pocket Dictionary. Springfield, MA: Merriam-Webster, 2006.

76. Moawd, R I, M A Talha, and REA El-Sharayhi. "Nutritional Evaluation of Sunflower Stalks Silage in Ruminants." Egyptian Journal of Animal Production 45, no. 1 (2008): 429–44. https://doi.org/10.21608/ejap.2008.104557.

77. Mohedano, Rodrigo A., Rejane H.R. Costa, Flávia A. Tavares, and Paulo Belli Filho. "High Nutrient Removal

Rate from Swine Wastes and Protein Biomass Production by Full-Scale Duckweed Ponds." Bioresource Technology 112 (2012): 98–104. https://doi.org/10.1016/j.biortech .2012.02.083.

78. Moroke, T.S., R.C. Schwartz, K.W. Brown, and A.S.R. Juo. "Water Use Efficiency of Dryland Cowpea, Sorghum and Sunflower under Reduced Tillage." Soil and Tillage Research 112, no. 1 (2011): 76–84. https://doi.org/10.1 016/j.still.2010.11.008.

79. Naik P.K., R.B. Dhuri and N.P. Singh, 2011. Technology for production and feeding of hydroponics green fodder. Extension folder No. 45/2011.

80. Naik P K, Dhuri R B, Swain B K and Singh N P. 2012. Nutrient changes with the growth of hydroponics fodder maize. Indian Journal of Animal Nutrition 29: 161–63.

81. Naik PK and Singh NP, 2013. Hydroponics fodder production: an alternative technology for sustainable livestock production against impending climate change. Model Training Course on Management Strategies for Sustainable Livestock Production against Impending Climate Change. Southern Regional Station, National Dairy Research Institute, Adugodi, Bengaluru, India Pp. 70-75.

82. Naik, P.K, R.B. Dhuri, M Karunakaran, B.K. Swain and N.P. Singh, 2014. Effect of feeding hydroponics maize

fodder on digestibility of nutrients and milk production in lactating cows. Indian Journal of Animal Sciences 84 (8): 880–883.

83. Naik, P. K., B. D. Dhawaskar, D. D. Faterpekar, B. K. Swain, E. B. Chakurkar, and N. P. Singh. "Yield and nutrient content of hydrophonics cowpea sprouts at various stages of growth." (2016a).

84. Naik, P.K., B.D. Dhawaskar, D.D. Fatarpekar, E.B . Chakurkar, B.K. Swain, and N.P. Singh. "Nutrient Changes with the Growth of Hydroponics Cowpea (Vigna Unguiculata) Sprouts." Indian Journal of Animal Nutrition 33, no. 3 (2016b): 357. https://doi.org/10.59 58/2231-6744.2016.00064.5.

85. Nelson, David L., Albert L. Lehninger, and Michael M. Cox. *Lehninger Principles of Biochemistry.* Macmillan International Higher Education: Basingstoke, 2021.

86. "New Zealand White Clover Cover Crop Seed." Johnny's Selected Seeds. Accessed 2022. https://www.johnnyseeds.com/farm-seed/legumes/clove rs/new-zealand-white-clover-cover-crop-seed-979.html.

87. Newell, Robert, Lenore Newman, Mathew Dickson, Bill Vanderkooi, Tim Fernback, and Charmaine White. "Hydroponic Fodder and Greenhouse Gas Emissions: A Potential Avenue for Climate Mitigation Strategy and Policy Development." FACETS 6 (2021): 334–57. https://doi

.org/10.1139/facets-2020-0066.

88. "News You Can Use: Aquaponics." h
ttps://www.hawaiinewsnow.com, February 12,
2012. https://www.hawaiinewsnow.com/story/169523
37/backyard-aquaponic-system-produces-green-greens/.

89. Nichols, Mike, and Wilson Lennard. "Aquaponics in
New Zealand." Practical hydroponics and Greenhouses
115 (2010): 46-51.

90. "Nutrition." Hydroponic Fodder Systems. Accessed August 2, 2022. http://foddertech.com/nutrition/.

91. Owade, Joshua O., George Abong,' Michael Okoth, and
Agnes W. Mwang'ombe. "A Review of the Contribution
of Cowpea Leaves to Food and Nutrition Security in East
Africa." Food Science & Nutrition 8, no. 1 (2019): 36–47.
https://doi.org/10.1002/fsn3.1337.

92. O'Callaghan, Tom F., Deirdre Hennessy, Stephen
McAuliffe, Kieran N. Kilcawley, Michael O'Donovan,
Pat Dillon, R.Paul Ross, and Catherine Stanton. "Effect
of Pasture versus Indoor Feeding Systems on Raw Milk
Composition and Quality over an Entire Lactation."
Journal of Dairy Science 99, no. 12 (2016): 9424–40.
https://doi.org/10.3168/jds.2016-10985.

93. Paolacci, Simona, Vlastimil Stejskal, Damien Toner, and
Marcel A.K. Jansen. "Wastewater Valorisation in an Integrated Multitrophic Aquaculture System; Assessing Nu-

trient Removal and Biomass Production by Duckweed Species." Environmental Pollution 302 (2022): 119059. https://doi.org/10.1016/j.envpol.2022.119059.

94. Peer, D.J., and S. Leeson. "Feeding Value of Hydroponically Sprouted Barley for Poultry and Pigs." Animal Feed Science and Technology 13, no. 3-4 (1985): 183–90. https://doi.org/10.1016/0377-8401(85)90021-5.

95. Plaza, Lucia, Begoña de Ancos, and Pilar M. Cano. "Nutritional and Health-Related Compounds in Sprouts and Seeds of Soybean (Glycine Max), Wheat (Triticum Aestivum.L) and Alfalfa (Medicago Sativa) Treated by a New Drying Method." European Food Research and Technology 216, no. 2 (2003): 138–44. https://doi.org/10.1007/s00217-002-0640-9.

96. "Population, Total." World. Accessed July 1, 2022. https://data.worldbank.org/indicator/SP.POP.TOTL?locations=1W.

97. Putt, Eric D. "Early History of Sunflower." Agronomy Monographs, 2015, 1–19. https://doi.org/10.2134/agronmonogr35.c1.

98. Rad, Fereidoon Haghighi, and Kavous Keshavarz. "Evaluation of the Nutritional Value of Sunflower Meal and the Possibility of Substitution of Sunflower Meal for Soybean Meal in Poultry Diets." Poultry Science 55, no. 5 (1976): 1757–65. https://doi.org/10.3382/ps.0551757.

99. "Reliable & Automated Animal Feed Technology." Hy-droGreen, May 17, 2022. https://hydrogreenglobal.com /.

100. Rep. North America Commercial Greenhouse Market Outlook 2027. Bonafide Research, 2022.

101. Resh, Howard M. Hydroponic Food Production. Mah-wah, Nj: New Concept Press, 2006.

102. Resh, Howard M. Hydroponic Food Production: A De-finitive Guidebook for the Advanced Home Gardener and the Commercial Hydroponic Grower. Boca Raton, FL: CRC Press, 2022.

103. S., Bakshi M P, and M. Wadhwa. "Hydroponics Green Fodder for Dairy Animals." Essay. In Recent Advances in Animal Nutrition. Delhi: Satish Serial Publishing House, 2014.

104. Salami, Saheed A., Giuseppe Luciano, Michael N. O'Grady, Luisa Biondi, Charles J. Newbold, Joseph P. Kerry, and Alessandro Priolo. "Sustainability of feeding plant by-products: A review of the implications for rumi-nant meat production." *Animal Feed Science and Tech-nology* 251 (2019): 37-55.

105. Saidi, Abd Rahim, and Jamal Abo Omar. "The Biological and Economical Feasibility of Feeding Barley Green Fod-der to Lactating Awassi Ewes." Open Journal of Animal Sciences 05, no. 02 (2015): 99–105. https://doi.org/10.

4236/ojas.2015.52012.

106. Schingoethe, David J., "Sunflower Seeds in Dairy Cattle Rations" (1992). Extension Extra. Paper 104. http://op enprairie.sdstate.edu/extension_extra/104

107. "Science: Hydroponics." TIME, March 1, 1937. http://content.time.com/time/subscriber/article /0,33009,757343,00.html.

108. Sedlak, David. Water 4.0: The Past, Present, and Future of the World's Most Vital Resource. Yale University Press, 2015.

109. Singh, B.B, H.A Ajeigbe, S.A Tarawali, S Fernandez-Rivera, and Musa Abubakar. "Improving the Production and Utilization of Cowpea as Food and Fodder." Field Crops Research 84, no. 1-2 (2003): 169–77. https ://doi.org/10.1016/s0378-4290(03)00148-5.

110. Singh S, Singh BS. Hydroponics - A technique for cultivation of vegetables and medicinal plants. In. Proceedings of 4th Global conference on Horticulture for Food, Nutrition and Livelihood Options, Bhubaneshwar, Odisha, India, 2012, 220p

111. Short, William A. Former USDA-FSIS Veterinarian on the Benefits of Feeding Fodder to Beef Cattle. Personal, September 16, 2022.

112. Sneath, R. and McIntosh, F., 2003. Review of Hydro-

ponic Fodder Production for Beef Cattle. Department of Primary Industries: Queensland Australia 84. McKeehen, p. 54.

113. Sommer, H., and A. Sundrum. "Determining the feeding value and digestibility of the leaf mass of alfalfa (Medicago sativa) and various types of clover." In Proceedings of 11th European IFSA Symposium, Farming Systems Facing Global Challenges: Capacities and Strategies, Berlin, pp. 1-4. 2014.

114. Sońta, Marcin, Anna Rekiel, and Martyna Batorska. "Use of Duckweed (Lemna L.) in Sustainable Livestock Production and Aquaculture – A Review." Annals of Animal Science 19, no. 2 (2019): 257–71. https://doi.org/10.24 78/aoas-2018-0048.

115. Stejskal, Vlastimil, Simona Paolacci, Damien Toner, and Marcel A.K. Jansen. "A Novel Multitrophic Concept for the Cultivation of Fish and Duckweed: A Technical Note." Journal of Cleaner Production 366 (2022): 132881. https://doi.org/10.1016/j.jclepro.2022.13288 1.

116. StoryMaps, ArcGIS. "(Farm) Animal Planet." ArcGIS StoryMaps. Esri, June 16, 2022. https://storymaps.arcgi s.com/stories/58ae71f58fd7418294f34c4f841895d8.

117. Swain, B.K., P.K. Naik, E.B. Chakurkar, and N.P. Singh. "Influence of Feeding Cowpea (Vigna Unguicu-

lata) Leaves on the Performance of Vanaraja Laying Hen."
Indian Journal of Animal Nutrition 35, no. 3 (2018): 371.
https://doi.org/10.5958/2231-6744.2018.00056.7.

118. Tahany, A. A., EL Rahim EA, S. A. Fayed, Amal M.
Ahmed, and M. M. F. Abdullah. "Influence of sprouting
on chemical composition and protein quality of radish
(Raphanus sativus) and clover (Trifolum alexandrinum)
seeds." J. Biol. Chem. Environ. Sci 13, no. 1 (2018):
339-355.

119. "The Complete History of Hydroponics and Water-Cul-
ture." Hydroponics, September 30, 2016. http://hydro
ponicgardening.com/history-of-hydroponics/.

120. Timko, Michael P., and B. B. Singh. "Cowpea, a multi-
functional legume." In Genomics of tropical crop plants,
pp. 227-258. Springer, New York, NY, 2008.

121. Tobey, Abigail. "Abigail's Fodder for Thought – Use
What You've Got!" FarmTek Blog, January 8, 2013.

122. https://farmtek.wordpress.com/2013/01/08/abigails-fo
dder-for-thought-use-what-youve-got/.

123. Tobey, Abigail. "Abigail's Fodder for Thought – Fall for
Fodder!" FarmTek Blog, September 5, 2013. https://far
mtek.wordpress.com/2013/09/05/fall_for_fodder/.

124. Torres, David E. Artist Depiction of The Hanging
Gardens of Babylon. HYDROPONICS: THE ART OF

GROWING PLANTS WITHOUT SOIL. Accessed February 2, 2021. https://www.landuum.com/en/history-and-culture/hydroponics-the-art-of-growing-plants-without-soil/.

125. "Total Water Use in the United States Completed." Total Water Use in the United States | U.S. Geological Survey. Accessed June 3, 2022. https://www.usgs.gov/special-topics/water-science-school/science/total-water-use-united-states.

126. "United States Department of Agriculture." USDA, September 2021. https://www.nass.usda.gov/Publications/.

127. "Urban Population (% of Total Population)." Data. Accessed July 2, 2022. https://data.worldbank.org/indicator/SP.URB.TOTL.IN.ZS.

128. US Sunflower Crop Quality Report. Mandan, ND: National Sunflower Association, 2021. https://www.sunflowernsa.com/uploads/38/2021SunflowerCropQualityReport.pdf

129. "Water Facts: UN-Water." UN. Accessed May 30, 2022. https://www.unwater.org/water-facts.

130. "Water in Agriculture." World Bank. Accessed June 30, 2022. https://www.worldbank.org/en/topic/water-in-agriculture.

131. "Water Uses: Colorado Water Knowledge: Colorado

State University." Water Uses | Colorado Water Knowledge | Colorado State University. Accessed June 30, 2022. https://waterknowledge.colostate.edu/water-manageme nt-administration/water-uses/#1525208065430-1ccb9fc 1-47bf.

132. "Water Week: How to Farm in a Drought." 1A, August 11, 2022. https://the1a.org/segments/water-week-how -to-farm-in-a-drought/.

133. White, Gary, and Matt Damon. The Worth of Water: Our Story of Chasing Solutions to the World's Greatest Challenge. Penguin Group USA, 2022.

134. Wood, Rommel, Manoush Zomorodi, Rachel Faulkner, and Katie Simon. "Ermias Kebreab: What Do Seaweed and Cow Burps Have to Do with Climate Change?" NPR. NPR, May 20, 2022. https://www.npr.org/2022/05/20/1099945356/ermias -kebreab-what-do-seaweed-and-cow-burps-have-to-do-w ith-climate-change.

135. Ziegler, P., K. Adelmann, S. Zimmer, C. Schmidt, and K. -J. Appenroth. "Relative in Vitro Growth Rates of Duck- weeds (Lemnaceae) - the Most Rapidly Growing Higher Plants." Plant Biology 17 (2014): 33–41. https://doi.or g/10.1111/plb.12184.